Plant Breeding

Plant Breeding

By

Manzoor Ahmad Khan

BIOTECH

2013

BIOTECH BOOKS

Delhi - 110 035

ISBN 81-7622-043-4
ISBN 978-81-7622-043-9

Published by : **BIOTECH BOOKS**
1123/74, Tri Nagar,
DELHI – 110 035
Phone: 27382765
e-mail: biotechbooks@yahoo.co.in

Showroom : 4762-63/23, Ansari Road, Darya Ganj,
NEW DELHI - 110 002
Phone: 23245578, 23244987

Printed at : **Chawla Offset Printers**
New Delhi – 110 052

PRINTED IN INDIA

CONTENTS

FIGURES

TABLES

FOREWORD

The need for locally-written textbooks on agriculture has been felt for a long time. This is because agriculture is a highly site-specific field, and students find it difficult to relate what is stated in textbooks written abroad to the widely different physical and socio-economic environment of Pakistan.

It became possible to meet this need with the obligation of USAID funds for this purpose under Project Implementation Letter No. 17 of 14 February 1990, largely due to the interest of Cordell Hatch, MART Project and Harry Dickherber, then of USAID, Islamabad. Strong interest and support were provided by former TIPAN Team Leaders Dr. Raymond G. Cragle and Dr. Gilbert H. Kroening, and by Abdur Rehman Khan, the former Vice Chancellor of the NWFP Agricultural University.

The project started with the constitution of a Task Force to plan the work. It comprised nominees of all the three agricultural universities, Pakistan Agricultural Research Council, National Book Foundation, USAID, and the TIPAN Project of the NWFP Agricultural University.

The first meeting of the Task Force was held on 25 March 1990. It identified the following topics and priorities for the textbooks to be written:

Textbook topic	**Priority**
Crop Production	1
Animal Production	1
Soil Science	1
Horticulture	1
Plant Breeding and Genetics	1
Extension Methods	1
Farm Management	2
Agricultural Entomology	2
Plant Pathology	2
Water Management	3
Range Management	3

It also identified managing authors and several contributing authors for the books at priority 1.

An authors' conference was held in Faisalabad on 6–8 May 1990, at which time panels of authors decided the contents of the books and named the most appropriate persons to write the various chapters. Dr. Everett Edington, Teaching Specialist, TIPAN Project, and Dr. J. Cordell Hatch,

Consultant to the MART Project, facilitated the conference. And thus the work got under way.

These textbooks have been written at the level of the beginning student of agriculture, and, as far as possible, cover indigenous research and experience.

The best part of the exercise has been the collaboration of over a hundred agricultural scientists from all over Pakistan.

S. Basit Ali Shah	Irshad Ali	Muhammad Rafiq Khan
Vice Chancellor	Vice Chancellor	Vice Chancellor
NWFP Agricultural University	Sindh Agricultural University	Agriculture University Faisalabad

Oval Myers, Jr.
Team Leader
TIPAN

EDITORS' PREFACE

Every effort has been made to present the ideas and interests of each of our authors as clearly and faithfully as possible, while at the same time adjusting the various formats of their separate articles into a consistent style suitable for a textbook. We have tried to keep the language as simple and straightforward as possible, consistent with accurate representation of the content. We hope that this will make the content of the book more accessible and useful to students.

Typographical conventions employed in this book are as follows. *Italic type* is used for: (1) botanical names at the genus and species level; (2) non-English words (e.g. Urdu, Russian, Latin, Pashto) which are not normally used in ordinary English conversation or writing (e.g. *in situ*, or *floris*); (3) titles of books mentioned in the text; (4) contrastive stress of a particular word or sentence (e.g. classified by *shape*, not function). Non-English words or botanical names which have become part of ordinary English usage in Pakistan or elsewhere are not italicized (e.g. berseem, dal, barani, rabi, kharif, chernozem, podzol, eucalyptus, delphinium, salvia). **Boldface type** is used to introduce an important new word or concept in the text.

Spelling generally follows the British tradition rather than the American. Individual author preference is respected, however, in this regard. Bibliographical conventions and abbreviations for standard units of measure are those specified in the fifth edition (1983) of the *CBE Style Manual*, published by the Council of Biology Editors, 9650 Rockville Pike, Bethesda, MD 20814, USA.

The editors take this opportunity to express their appreciation to the authors for the spirit of cooperation and enthusiasm which they have shown at all stages of our endeavor to produce the best possible locally-oriented textbook. The illustrations in the book would not have been possible without the invaluable help of Professor Habib-ul-Rahman Mian of the Department of Plant Breeding and Genetics, NWFP Agricultural University—both in the coordination and execution of the artwork. We also wish to thank Mr. Ataullah, Librarian, and Mr. Rahim Dad, Assistant Librarian, NWFP Agricultural University, for their help in finding hard-to-locate materials in the library, and all the other members of the NWFP Agricultural University faculty with whom we have worked. We feel that during the course of this project we have become part of the NWFP AU institutional family.

INTRODUCTION

In their earliest incarnation, *Homo sapiens* must have been gatherers of food, mostly of plant origin. In this process, they must have recognized plants which were of particular value to them. This must have been followed by the domestication of particularly useful plants, and much later, by their conscious breeding. Manzoor Ahmad Khan and Iftikhar Ahmad Khan give a brief account of the early history of plant breeding, and mention a few disciplines of biology related to plant breeding.

Habib-ul-Rahman Mian in his chapter on modes of reproduction, starts laying the foundation which is needed for understanding the subject of plant breeding. Since conventional plant breeding depends on the process of sexual reproduction, Habib-ul-Rahman describes the parts of the flower, and the processes of pollination and sexual reproduction; wherever required, genetic and cytological explanations are also given. Asexual reproduction is briefly described towards the close of the chapter. The chapter introduces several terms which it would be well to understand thoroughly and to remember by heart, because they comprise the language of the plant breeder.

Khurshid Alam's "Breeding Methods for Self-Pollinated Crops" introduces the basic methods of plant improvement: introduction, pure-line selection, mass selection, and hybridization. The bulk of the chapter is devoted to hybridization, in which he describes how to make a cross and how to process its progeny through the pedigree and bulk methods. The last section of the chapter deals with the backcross method of varietal improvement, which has been used extensively in wheat for incorporating disease resistance into commercial varieties of wheat and other field crops.

Muhammad Saleem and Hidayat-ur-Rahman cover the methods of breeding cross-pollinated crops. Included are: selection, hybridization, and the development of synthetic varieties. Most of the examples of the application of breeding procedures presented in the chapter concern maize, because it is one of the most important cross-pollinated species of Pakistan, and also the subject of Saleem's own work.

Improvement of crops which are normally propagated by asexual means is described by Muhammad Siraj Swati and Hidayat-ur-Rahman. Topics covered comprise the characteristics of clones, advantages and disadvantages of asexual propagation, sources of genetic variation within clones, causes of clonal degeneration, and methods of improvement of vegetatively propagated crops: introduction, selection, sexual methods, and recent techniques of mutation breeding and tissue culture. This is followed by an enumeration of the problems inherent in the breeding of asexually propagated crops.

New advances in plant breeding are covered by one of their foremost practitioners in Pakistan, Khushnood Siddiqui. Topics covered include mutation breeding, new advances in plant cytogenetics, and *in vitro* techniques in plant breeding. Under *in vitro* techniques are described such state-of-the-art methods as tissue culture, cell-suspension culture, protoplast culture, anther culture, *in vitro* culture of ovules, *in vitro* pollination and fertilization, embryo rescue, micropropagation, and new biochemical and molecular inputs in plant breeding. The chapter is heavy reading but brings a breath of fresh air to this discourse; once mastered, it puts you at the forefront of contemporary plant breeding.

A knowledge of experimental design and analysis is extremely important in plant breeding. It enables the breeder to assess whether the superior performance of a new cultivar over an established one is really due to its hereditary traits or can be explained by chance variation, which always occurs when the same experiment is repeated under field conditions. This important subject is covered by Syed Sadaqat Mehdi and Iftikhar Ahmad Khan. Topics covered are measures of variability, the completely randomized and randomized complete block experimental designs, and methods of estimating heritability.

Investment in plant breeding research and the development of crop varieties is fully repaid only if the seed of desirable varieties can be multiplied and distributed to growers without loss of quality, at reasonable rates. Abdul Majid Chaudhry attempts to describe the system prevalent in Pakistan. The main components of the chapter are a description of seed quality, and of Government seed quality control agencies. At the end of the chapter are appended the minimum seed certification standards for various crops prevalent in Pakistan, and the seed sampling rules promulgated by the Government of Pakistan in 1991.

Continuous genetic improvement is central to the task of increasing agricultural productivity. No amount of husbandry can make a crop cultivar perform beyond its genetic potential. Advances in plant breeding and genetics are not only continuously increasing crop yields but are also providing mechanisms for incorporating into crop plants tolerance against pests and diseases and physiological stresses caused by adverse climatic and soil conditions. Such spectacular progress in plant breeding and genetics has been made possible by the variety of genetic resources available to plant breeders in the numerous land races of crops, the progenitors and wild relatives of crop plants, and the new germ plasm being created through the application of modern techniques. But the genetic resources of the world are shrinking.

Most land races are going out of cultivation in favour of a narrow range of modern cultivars. And the progenitors and wild relatives of crop plants are being exterminated by the overgrazing of natural vegetation. A. S. Larik discusses this subject in his chapter on genetic resources and their conservation, and points the way to maintaining and expanding this vital resource for the present and future prosperity of the human race.

Plant breeding has been one of the most active fields of agricultural science in Pakistan. Muhammad Amin Khan and Iftikhar Ahmad Khan, in their chapter on "Improved Varieties of Crop Plants in Pakistan", describe several cultivars developed by our crop breeders in wheat, maize, sugarcane, oilseeds, cotton, pulses, and fodder. Muhammad Saleem supplements this account with information for NWFP, particularly in maize, which is his specialty.

G. M. Khattak
Coordinator

Manzoor Ahmad Khan
Managing Author

TO THE STUDENT:

This textbook has been written by a team of agricultural researchers and teachers from various parts of Pakistan, each with his own particular expertise and focus of interest. In order to make its contents maximally accessible to you, its users, we have provided the following features.

- **Chapter outlines.** Each chapter is preceded by an outline of its main content areas. Read this first to get a broad idea of the structure of the chapter and the topics it covers.
- **Learning objectives.** At the beginning of each chapter is a list of learning objectives; that is, things which you should be able to do after studying the chapter. Read and think about these before you begin to study the chapter to clarify in your mind what you are going to learn about. Try to keep them in mind to guide and focus your reading of the chapter.
- **Study questions.** At the end of each chapter there is a set of questions. These are intended to re-focus your attention on the most important content of the chapter. They can be used for self-study or assigned by your teachers.
- **References.** A list of references follows each chapter. This list contains all works referred to in the text, as well as suggestions for further reading in some cases. This is a valuable tool for you to use in pursuing an idea beyond what is given in the textbook itself.
- **Abbreviations.** Abbreviations and acronyms used specifically in this book as well as those for standard units of measurements are listed in a separate section at the end of the book.
- **Glossary.** After the final chapter there is a glossary of important technical terms and words which the authors have felt may need to be precisely defined. Use this like a mini-dictionary to find the meaning of an unfamiliar or difficult word quickly.
- **Index.** At the end of the book there is a topic index. In it are listed words and phrases which refer to important concepts, institutions, or persons. If the topic or idea you are interested in is not listed in the table of contents, look for it in the index.

Outline of Chapter 1

1. HISTORY OF PLANT BREEDING

Manzoor Ahmad Khan
Iftikhar Ahmad Khan[1]

LEARNING OBJECTIVES

After reading this chapter, a student should be able to:

- State the major difference between early attempts at improving crops and present-day plant breeding.
- Explain the basic idea of Lamarckism.
- Discuss Mendel's work and outline the basic ideas he contributed.
- Discuss some of the major advances in the understanding of wheat genetics, and the uses plant breeders have made of them.
- Suggest a program of specific goals for plant breeders in Pakistan.

1.1 Definition of plant breeding

From the time man began to form settlements rather than pursue a nomadic existence, the plants surrounding him took on greater importance. Civilizations were located in areas where wild food plants grew abundantly. These plants were diverse in form and productivity, and the first plant breeding undoubtedly consisted of saving seed from the most desirable plants to plant in the next season. Such primitive selection was the basis for what is commonly called plant domestication or the modification of plants for the use of mankind.

Simmonds (1976) lists over 80 major crop plants along with details of their early and recent domestication. They include wheat, maize, barley, beans, sorghum, cotton, rice, millet, soybean, sugarcane, rubber, tea, tobacco, beet, potato, onion, apple, and citrus. The selection of these plants for domestication consciously or unconsciously was the result of need and depended upon the artistic skills of man in finding chance variation, since it preceded any biological knowledge based on scientific principles. After the

---

[1]Dr. Manzoor Ahmad Khan is Dean, Faculty of Agriculture, University of Agriculture, Faisalabad. Dr. Iftikhar Ahmad Khan is in the Department of Plant Breeding and Genetics, University of Agriculture, Faisalabad.

discovery of the Mendelian laws dealing with the inheritance of genetic variability, the work of improving plants and animals was put on a scientific basis. Now the process of plant improvement has become more of a science than a simple art. Against this background, plant breeding may be defined as the *art and science of changing and improving the genetic patterns of crop plants in relation to their economic use, or the application of genetic principles for improving plants in relation to their economic use.*

1.2 Pre-Mendelian and modern concepts of plant breeding

Although the history of plant breeding is linked with the history of mankind as he tried to select, domesticate, and improve plants as a source of food and fiber, it is difficult to establish when man consciously began to breed plants. Certainly, considerable plant improvement occurred prior to a knowledge of genetics. As mentioned earlier, modern plant breeding means the improvement of plants through the application of genetic principles. Therefore the historical development of important genetic concepts is also a part of the history of plant breeding. Some of the important events and concepts which were developed before 1900 when Mendel's work was rediscovered and the foundations of modern genetics and plant breeding were laid, are described here in brief as pre-Mendelian concepts of plant breeding.

Artificial pollination of date palms was practiced by the Assyrians and Babylonians as early as 700 B.C. The native Americans were successful in breeding maize long before the western hemisphere was discovered by the Europeans. A large number of early workers contributed to the evolution of the science of plant breeding, but an experimental approach to plant improvement through hybridization was first made possible by the work of R.J. Camerarius (1665–1721), a German professor of medicine. He reported the existence of sexual reproduction in flowering plants. He showed that in maize, seeds are not produced unless pollen is applied to the pistils, and concluded that the pollen was the male and the pistil the female element. He is also credited with the first artificially produced plant hybrid. Later J.G. Kolreuter (1733–1806) performed experiments on artificial hybridization of tobacco varieties and studied their hybrids. He provided clear evidence that in reciprocal crosses, the hereditary contribution of the two parents to their offspring was equal.

C. de Lamarck (1744–1829) in 1809 theorized that species exposed to natural or artificial environmental conditions can change gradually into new ones through a constant strengthening and perfecting of adaptive characteristics and that the acquired characteristics are transmitted to the offspring. This theory was called **Lamarckism.** This false theory of the inheritance of environmentally acquired characters was later advocated by Lysenko in the former Soviet Union during the early years of communism and delayed progress in plant breeding in that country for several years.

Some important developments took place during the first half of the 19th century. For example, in 1828 R. Brown discovered the cell nucleus in the flowering plant *Tradescantia,* and in 1835 H. Von Mohl described cell division. M.J. Schleiden and T. Schwan in 1839 proposed the cell theory, stating the cell to be the unit of biological organization; and this concept was immediately and universally accepted. Gordron in 1844 concluded that sterile hybrids were due to distinct parental species and that only crosses of two races within a species gave fertile hybrids.

During the period from 1850 to 1900 there developed improved understanding of the most important biological phenomenon, the concept of heredity and the basis of genetic variation. Darwin's work on the origin of species by means of natural selection appeared in 1859. His hybridization studies and development and elaboration of the concepts of selection and evolution dominated the period up to 1900. Many workers of the period were stimulated by Darwin's work and conclusions. F. Galton developed the concept of regression, a measurement of the degree of resemblance of parents, progeny, and relations.

The idea of the union of sperm and egg as constituting the process of fertilization was the conclusion of O. Hertwig in his 1877 studies on sea urchins. In 1882, W. Flemming proposed the term **mitosis** and explained the behavior of chromosomes during this process. He also applied the name **chromatin** to the stainable portion of the nucleus.

In 1884, E. Strasburger described fertilization in angiosperms. He also made reciprocal crosses between different plant species and found that the results were similar to those of direct crosses. T. Boveri discovered the real nature of reduction division and described **meiosis** and, particularly, synapsis (pairing of homologous chromosomes) in *Ascaris.*

August Weismann in essays of 1883 and 1885 put forth his **germ plasm theory**, which was an alternative explanation to Lamarck's theory of acquired characteristics. He believed that the reproductive tissue (germ plasm) is quite separate and distinct from the body tissue (somatoplasm). He emphasized the remarkable stability of germ plasm which is transmitted unchanged from generation to generation. Somatoplasm is the product of germ plasm, and, therefore, any change induced by the environment on the body cells cannot be transmitted to the germinal tissue. Acquired characters are not inherited because germ plasm is the only channel through which heredity is carried to the offspring.

Work on plant improvement, particularly through selection and hybridization, also continued during this period. For example, in 1857 Goodrich produced an improved variety of potato through selection. Henry and de

Vilmorin described the interspecific wheat crosses, *Triticum aestivum* × *T. turgidum* and *T. durum* × *T. polinicum* in 1878. Beal (1878–81) found increased yield from hybrids between varieties, and suggested use of varietal crosses in maize production. Similarly, Blount (1879) produced inbreds in maize by controlled hand pollination, and Harsford (1879–80) selected the first hybrid variety of barley in the USA. In 1899, Hopkins described the ear-to-row method for the improvement of maize in the USA.

The 20th century brought the science of genetics into plant breeding with the rediscovery in 1900 of the Mendelian laws of inheritance. In fact, Mendel had presented his work before the History Society of Brunn in 1865 and his paper on "Experiments in Plant Hybridization" was published in 1866, but this work remained unnoticed for 34 years. It was rediscovered and given due credit in 1900 when three biologists, J. Correns, H. De Vries, and E. Von Tschermak read Mendel's paper and showed that they had also obtained similar results about the fundamental principles of heredity from their own experiments. The new principles became known as Mendel's Laws: the Law of Segregation and the Law of Independent Assortment.

In simple terms the **Law of Segregation** states that the factors responsible for a trait are contributed equally by each parent to the offspring and that the offspring will segregate (or pass on) only one parent's contribution to their own individual offspring. This was opposed to the concept of blending inheritance prevalent in the late 19th century. The **Law of Independent Assortment** states that the factors responsible for two separate traits segregate independently (or separately) from each other when the parent passes these factors on to their offspring. This Law is very important for plant breeding in that it permits new combinations of traits. Thus a hybrid offspring resulting from a cross between a tall, white-flowered parent and a short, red-flowered parent might produce offspring that were tall and red-flowered or short and white-flowered.

As mentioned earlier, man's efforts to improve his plants could be called plant breeding in a modest way but there was no science in it; it was simply an art of selection. Plant breeding continued to be an art up until the end of the 19th century. Modern techniques of crop improvement developed on a sound scientific basis only after the rediscovery of the Mendelian laws of inheritance.

Rediscovery of Mendel's work brought a breakthrough in biology and particularly significant developments occurred in the fields of genetics, cytology, biometry, and evolution—the basis of plant breeding. Some of the important contributions made in these fields after Mendel are as follows De Vries in 1902 contributed the mutation theory of evolution. William Bateson and R.C. Punnett (1902–07) verified and extended the Mendelian principles. They also reported the first case of linkage in sweet peas. Linkage is an exception to Mendel's Law of Independent Assortment in that factors responsible for two different traits do not separate independently because they are on the same large unit of inheritance, the chromosome. It should be noted that chromosomes and the processes of mitosis and meiosis were discovered after Mendel proposed his theories of inheritance. It was probably fortunate for Mendel and for genetics and plant breeding that the traits Mendel studied were independently inherited.

In 1903, W.L. Johannsen developed the pure line theory and in 1905 coined the terms, **gene, genotype,** and **phenotype**. Sex linkage in the *Drosophila* fruit fly was discovered by T. M. Morgan in 1910. In 1916, Wrinkler introduced the term **genome** for a set of chromosomes. Shull in 1916 suggested using a single cross to increase production in maize, and named the increased vigor or production of the cross over the mean of the parents **heterosis.**

The theory of polyploid origin of plant species by multiplication of genomes was proposed by Winge in 1917. Sakamura in 1918 proposed classification of wheat species on the basis of chromosome numbers. Kihara in 1919–24 reported that two genomes of hexaploid wheat are equivalent to those of tetraploid wheats. Hayes et al. (1920) transferred genes of durum to bread wheat for stem rust resistance.

People working in the area of field experimentation and biometry helped the field of plant breeding in its development. For example, Wood and Stratton in 1910 suggested the use of check plots, and Mercer and Hall introduced the concept of replication in field experiments in 1911. Harris suggested the use of chi-square analysis in genetic experiments in 1912. The use of biometry or statistics in plant breeding has become increasingly useful as complex quantitative traits due to many genes can best be studied utilizing statistics to supplement genetic understanding.

During the third and fourth decades of the 20th century there was a rapid development of knowledge and procedures in disciplines related to plant breeding. In 1921, Blakeslee gave the term **monosomic** to plants deficient by one chromosome. While studying the cross *Triticum monococcum* × *T. turgidum*, Sax (1922) suggested that the A-genome of wheat might be related to a diploid species; Percival in 1926 explained the polyploid origin of today's tetraploid and hexaploid wheats. A very useful discovery, the cytoplasmic male sterility of maize, was made by Rhoades (1933). It contributed a lot in producing hybrids in maize, sorghum, wheat, and other crop plants. Working on similar questions, Nishiyama (1929) reported extensive work on the cytogenetics of oat, and Sears (1939) on that of wheat and its relatives. Today's mutation breeding was introduced in 1928 by Stadler when he described the mutagenic effects of X-rays in barley. The use of colchicine discovered by Dustin in 1934 was demonstrated by many studies to double chromosome numbers and allow production of polyploid plants. Production of polyploids has been most useful in horticultural plants and in studies of crop plant evolution.

Significant enhancement occurred also in biometry during this period. A series of papers on mating systems were published. The first of those was that of Wright in 1921, "The Biometrical Relations Between Parent and Offspring" (Wright 1968). Fisher introduced a methodology of analysis of variance in 1925 which is still in wide use by plant breeders. Experimental designs, analysis of variance methods, and proper field plot techniques were explained comprehensively by Love in 1936 and Snedecor in the mid-1930's.

After the fourth decade of the 20th century, the pace of the research work in plant breeding accelerated, helped by the parallel development of many helping sciences and technical improvements. The establishment of research organizations, institutes, and communication of research results among plant breeders and other scientists through summaries, work of conferences, and symposia have been most effective in making information from related sciences available to plant breeders.

Over the period 1948–56, Sears reviewed and summarized cytogenetic work on wheat. Kihara (1956) reviewed the classification of some of the wheat genera. Similar summaries were supplied for barley workers by Smith (1951) and Nilan (1964). Rice species and their genetic behavior were described by Nagao (1951) and Morinaga (1956). Peterson (1958) described wheat breeding history and progress in Canada. Klose (1950) and Hodgson (1961) wrote reviews of plant introduction and the uses made of germ plasm sources. Jinks and Hayman in the period 1954–56 developed the diallel cross technique for the genetic analysis of different agricultural crops (Jinks 1970).

In the early 1960's, the introduction of dwarfism in crops like wheat, rice, sorghum, maize, and soybean was accomplished. This ultimately resulted in higher production per unit area and helped significantly in bringing the Green Revolution to the whole world.

All the processes involved in the improvement of plants and animals through the application of genetics principles and other conventional techniques are included in the domain of biotechnology. But today's biotechnology has been enriched by new concepts like genetic engineering and somatic cell genetics. New knowledge and techniques arising from research in the biological sciences are increasingly leading to novel ideas for improving the productivity of crops and livestock. Some people have claimed, for example, that advances in biotechnology will give rise to a second Green Revolution, that crops will be engineered to flourish in marginal environments, and that traditional methods of plant breeding will rapidly become obsolete.

1.3 Disciplines related to plant breeding

A plant breeder has to continually adjust plants genetically to the climatic, social, cultural, economic, and technological aspects of the environment. It is essential that he know these plants in detail. Knowledge of genetics underlies plant breeding methods. Apart from genetics, many other disciplines deal directly or indirectly with plant breeding.[2]

Botany. A plant breeder should understand the morphology, taxonomy, and reproduction of the plants with which he works.

Evolution. Plant breeding may be considered a form of directed evolution, and for this reason the plant breeder should be thoroughly familiar with patterns of evolution in natural species.

Cytogenetics. In addition to an understanding of the mechanisms of heredity, a plant breeder should also be knowledgeable about chromosome behavior and cytoplasmic inheritance.

Plant physiology. A knowledge of plant physiology helps the plant breeder to understand the effect of environmental factors on plant growth and to modify plants for growth under such conditions.

Plant biochemistry. Nutritional value and suitability for industrial use often determine the market demand for a particular variety or crop. Cotton seed, for example, is an important source of edible oil in Pakistan. The studies and tests needed to eliminate the gossypol (a poisonous alkaloid) present in the seed require knowledge of chemistry. Similarly, the milling and baking qualities of wheat, the cooking qualities of rice, and the nutritional value of other crops require a basic knowledge of biochemistry. Along with its role in better understanding mutation and gene action, a knowledge of biochemistry is necessary for successful exploitation of modern biotechnology in the field of plant breeding.

Plant pathology. Many fungal and bacterial diseases affect crop plants. For example, rust is a problem in wheat and bacterial blight in cotton. For a productive crop, the variety must have resistance against major plant diseases. To develop a resistant variety, the breeder must have an understanding of plant pathology.

Entomology. Insect pests play havoc with our agricultural crops. Some of the major crops in Pakistan, e.g. cotton, rice, and sugarcane, are subject to a variety of insect pests in the field. A plant breeder, therefore, must have a knowledge of entomology to successfully breed for varieties resistant against the prevailing insect pests.

---

[2]The discussion in this section follows Poehlman (1987:4).

Agronomy. Since experimental lines, breeding populations, and materials should be grown using standard farming techniques, a knowledge of crop growing practices is essential for a successful plant breeder.

Statistics. A plant breeder has to evaluate a large number of advanced breeding lines and predict their performance. Good field-plot techniques and statistical analyses are necessary to obtain reliable data. The understanding of quantitative inheritance is also based on a knowledge of statistical procedures or the biometrical approach to plant breeding.

A plant breeder does not need to be an expert in all the above fields of science, but should be able to apply the knowledge and experience of these disciplines in developing varieties. Research and advances in these disciplines are continually integrated into plant breeding procedures to make them more efficient and to facilitate varietal improvement. Apart from knowledge of scientific disciplines, a plant breeder must constantly keep in mind those vitally important socio-economic considerations that often determine the success of new varieties.

1.4 Objectives of plant breeding

For the rapidly growing world population, a continually increasing supply of basic foodstuffs is needed. In Pakistan, where the rate of population increase is one of the highest in the world and food supply is inadequate, the situation is alarming. It is, therefore, essential to give serious consideration to the problem of food increase in Pakistan to avoid a worse situation in the years ahead. This problem will not be solved only by increasing the quantity of food; improvement in quality and nutritional level is also needed.

Five major factors must be given major attention to increase crop production: soil, water, fertilizer, pest control, and crop variety. The first four concern providing an environment for crop growth. The fifth, crop variety, relates to the inherent ability of the plant to produce within the existing environment. Maximum crop production can be achieved by improving both the crop variety and the environment in which it is to grow.

The goal of plant breeders is to provide varieties which are better than the existing ones. Higher yield is and will remain the foremost goal; however, disease and insect resistance, improvement in quality in respect of taste, nutrition, industrial use, and keeping quality of fruits, vegetables, flowers, and food grains are all important.

Germ plasm resources. Plant breeders in Pakistan should try to accumulate a collection of diverse genotypes which may be used as source material for desirable genes. The collection can include both local and exotic strains. The seed or plant stocks thus collected must be catalogued, maintained in viable form, and shared with plant breeders within Pakistan.

Short-duration varieties. Early-maturing and short-duration varieties may save a lot of time, labor, and other expenses of the farmer. In the crop growing system of Pakistan, where wheat follows rice or cotton, such varieties are especially helpful to the farmers to harvest and sow their crops satisfactorily. Early-maturing varieties also escape the onslaught of various pests. Plant breeders should, therefore, strive to breed for such varieties in our agricultural crops.

Stress-resistant varieties. Fitting crops to environments is as old as plant culture in agriculture. The agricultural tracts of Pakistan have a variety of stressful conditions that our crops have to face. A large cultivable area is *barani* (rainfed) and has moisture stress; salinity and waterlogging are severe problems; also, hot seasons in central Punjab affect our crops badly. All these factors collectively cause a significant decrease in our total agricultural production. Although much progress has been made in adjusting our crops to some of these situations, much more can be accomplished if breeders and agronomists cooperate to identify adverse environments and breed plants more tolerant of such conditions.

Disease and insect pest resistance. Exploitation and improvement of host-plant resistance against prevalent disease and insect pests is one of the major activities of plant breeders in all countries working on almost all crops. Since this is the cheapest and the safest way to protect crops, plant breeders of Pakistan should try to develop plants which are resistant or will synthesize their own pesticides. Some of the more important biotechnology uses may be in this area.

Oilseed crops. Because of her meager oilseed production, Pakistan has to import three-fourths of her edible oil requirements at the cost of a huge amount of foreign exchange. Apart from improving the conventional oilseed crops in Pakistan, breeders should also explore new sources of edible oil. More than 55% of the oil production of Pakistan is contributed by cotton. The oil content of cottonseed should be improved to help solve the problem of edible oil.

Fertilizer-responsive varieties. Plant breeders should try to genetically engineer varieties which are highly responsive to fertilizers and can mobilize their dry matter producing capacity for increased harvest index. For example, introduction of short-strawed or semi-dwarf wheats that are highly responsive to heavy application of fertilizer has boosted the national average yield in Pakistan. Such sources of dwarfing genes must be explored and exploited in other crops as well.

Hybrid varieties. Hybrid production in outbred crop plants provides a means of fulfilling a breeder's aim of high yield and uniformity of field performance and quality. Efforts should also be made to extend the exploitation of hybrid vigor or heterosis on a commercial scale in other crops like wheat, rice, and cotton.

Adaptation to mechanical harvesting. Farm mechanization is being rapidly adopted in Pakistan. Our crop varieties should, therefore, be made suitable for the use of machines for various field operations. For example, in cotton, a small, compact plant which does not lodge and with bolls set high off the ground is best suited for a picker-type harvester.

Plant genetic engineering. For a plant breeder, plant genetic engineering has the potential to transfer a gene for a beneficial character from a distantly related wild species into an economical species when sexual hybridization is not possible. Plant breeders in Pakistan should also try to develop the necessary knowledge and infrastructure for the application of this technique to engineer future plants.

Seed multiplication. Although seed multiplication is only partly the role of the plant breeder, it is critical to the ultimate value of new crop varieties or cultivars. In developed countries, seed multiplication is rarely a problem since the private seed sector is well organized; however, for countries like Pakistan where the seed industry is poorly developed, the plant breeder needs to take a more active role in the multiplication and dissemination of new cultivars.

STUDY QUESTIONS

1. Define plant breeding.
2. Why is the process of plant improvement before the rediscovery of Mendelian Laws considered to be an art rather than a science?
3. Explain how various scientific disciplines help in the process of plant breeding.
4. What should our breeders do to meet the challenge of population increase in Pakistan?

REFERENCES

Arnold, M. Biotechnology creates opportunities but poses problems too. Ceres: The FAO Review 21(3):21–24; 1988.

Chaudhary, R. C. Introduction to plant breeding. New Delhi: Oxford and IBH Publishers; 1989.

Frey, K. J. ed. Plant Breeding: A symposium held at Iowa State University. Ames, Iowa: The Iowa State Univ. Press; 1966.

Hayes, H. K.; Parker, J. H.; Kurtweil, C. Genetics of rust resistance in crosses of *Triticum vulgare* with varieties of *T. durum* and *T. cicoccum*. J. Agric. Res. 19:523–542; 1920.

Hodgson, R. C. ed. Germ plasm resources. Pub. 66. Washington, D. C.: American Association for the Advancement of Science; 1961.

Jinks, J. L. ed. Fifth years of genetics. Edinburgh, Scotland: Oliver and Boyd; 1970.

Kihara, H. Wheat and its relatives. Reprint from Exhibits, Intl. Genet. Symp., Tokyo. Kyoto: Sci. Council of Japan; 1956.

Klose, N. America's crop heritage. Ames, Iowa: Iowa State College Press; 1950.

Mayo, O. The theory of plant breeding. 2nd ed. Oxford: Clarendon Press; 1987.

Morinaga, T. The rice. Reprint from Exhibits, Intl. Genet. Symp., Tokyo. Kyoto: Sci. Council of Japan; 1956.

Nagao, S. Genic analysis and linkage relationship of characters in rice. Advan. Genet. 4:181–212; 1951.

Nilan, R. A., The cytology and genetics of barley. Res. Stud., Monographic Suppl. No. 3(32):1–278. Pullman, WA: Washington State University; 1964.

Nishiyama, I. The genetics and cytology of certain cereals. I. Morphological and cytological studies on triploid, pentaploid and hexaploid Avena hybrids. Jap. J. Genet. 5:1–48; 1929.

Peterson, R. F. Twenty-five years' progress in breeding new varieties of wheat for Canada. Emp. J. Exp. Agr. 16:104–122; 1958.

Poehlman, J. M. Breeding field crops. New York: Avi Publishers; 1987.

Poehlman, J. M. and Borthakur, Dh. Breeding Asian field crops. New Delhi: Oxford and IBH Publishing Co.; 1959.

Rhoades, M. M. The cytoplasmic inheritance of male sterility in *Zea mays*. J. Genet. 27:71–93; 1933.

Sax, K. Sterility in wheat hybrids. II. Chromosome behavior in partially sterile hybrids. Genet. 7:513–552; 1922.

Schulz–Schaeffer, J. Cytogenetics: plants, animals, humans. New York: Springer–Verlag; 1980.

Sears, E. R. Cytogenetic studies with polyploid species of wheat. I. Chromosomal aberrations in the progeny of a haploid of *Triticum vulgare*. Genet. 24:509–523; 1939.

Simmonds, N. W. Evolution of crop plants. London: Longman; 1976.

Smith, L. Cytology and genetics of barley. Botan. Rev. 17:1–51, 133–202, 285–355; 1951.

Wright, S. Evolution and the genetics of populations, I. Chicago: University of Chicago Press; 1968.

Outline of Chapter 2

2. MODES OF REPRODUCTION IN CROP PLANTS

Habib-ul-Rahman Mian[1]

LEARNING OBJECTIVES

After reading this chapter, a student should be able to do the following things.

- Use appropriately the specific terminology which describes the various reproductive mechanisms in crop species.
- Explain essential features of plant reproduction and discuss how these features influence procedures used by plant breeders.
- Discuss the different morphological, functional, and genetic phenomena that affect and modify the reproductive behaviour of plants.
- Explain how the specific reproductive mode of a plant species influences its breeding behaviour and the choice of breeding method appropriate for it.
- Discuss the applications of specific knowledge of the reproductive behaviour of plant species to other related topics of plant science in general and plant breeding in particular.

2.1 Introduction

Generation after generation, the continuity of life is maintained through reproduction. The most important distinguishing feature of living organisms is the ability of every species to reproduce. Reproduction is a means of increasing the population as well as maintaining the continuity of a species. In addition to this, it plays an important role in other aspects of life. (i) Plants generally depend on dispersal of reproductive structures such as seeds, spores, or even vegetative parts for movement from one favourable habitat to another. (ii) Hereditary variations are transmitted from parents to offspring according to the principles of genetics, which leads to the

---

[1]Dr. Habib-ul-Rahman Mian is Associate Professor, Department of Plant Breeding and Genetics, NWFP Agricultural University, Peshawar.

evolution of new and diverse races, varieties, or species. For most crop species, the modes of reproduction are now well understood; however, when breeding work is initiated on new crop varieties or species, a thorough acquaintance with the plant's reproductive mechanism becomes essential for the individual plant breeder. Reproduction in plants can occur by three primary methods.

i. Sexual means, in which two haploid gametes fuse by a process called fertilization to form a diploid zygote that eventually develops into an embryo and seed. The seed then grows into a subsequent generation. As a rule (however, with a few exceptions), the members of this new generation may genetically differ among themselves as well as from the parents.

ii. Asexual means of reproduction involve the rather complicated process of **apomixis** or **agamospermy**. This essentially signifies the production of vegetative plant parts from plant organs that are basically associated with sexual reproduction. It is a phenomenon of seed production in which the embryo develops from the cells of the female gametophyte alone, due to various disruptions in the normal sexual process. This will be discussed in more detail in subsequent pages. Figure 2.1 schematizes the processes of sexual and asexual reproduction.

iii. Vegetative reproduction, in which new plants develop from somatic tissue of the parent plant. Vegetative parts like stem cuttings, root cuttings, bulbs, tubers, or even single cells are commonly used. Plants that develop by vegetative reproduction will be genetically identical to the parents. If a large population is raised by stem cuttings obtained from a single parental plant, all the members in this population will be genetically identical and called a **clone**. Vegetative reproduction is, therefore, often also referred to as **clonal propagation**.

2.2 The flower

Flowers are highly specialized organs of angiosperms in which sexual reproduction and eventually seed and fruit formation occur. The range of pattern among the flowers of various species is vast, and their forms, sizes, and composition are infinitely varied. We can only touch here on the basic features and general characteristics of a typical flower from a breeder's point of view. The parts of a typical flower are shown in Figure 2.2.

Botanically, all the parts of a flower are extremely modified leaves, arranged in successive whorls on the **thalamus** or receptacle which is an enlarged end of flower stalk or **pedicel**. In the very young or bud stage, the flower is completely enclosed in two sets of leafy whorls. The outermost whorl, called the **sepals**, encloses and protects all the other parts. The sepals are often small, green, and tough, but in some species they are large or

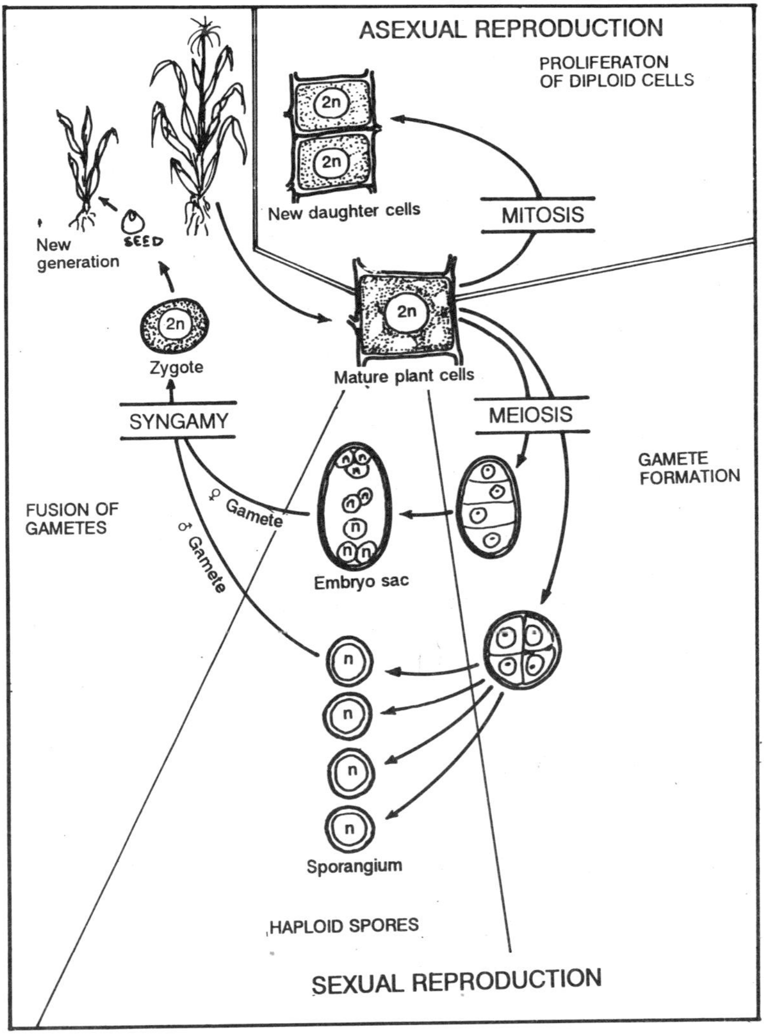

Figure 2.1 A basic scheme showing the reproductive process in plants. Asexual reproduction is a direct outcome of mitosis, whereas sexual reproduction occurs with the formation and fusion of gametes.

variously coloured. Every species has a specific form and number of sepals. They may develop as free leafy structures or become joined to form a cup or tubular structure. All of them together are called the **calyx**. An inner whorl of sepals, collectively called the **corolla**, is often the most conspicuous,

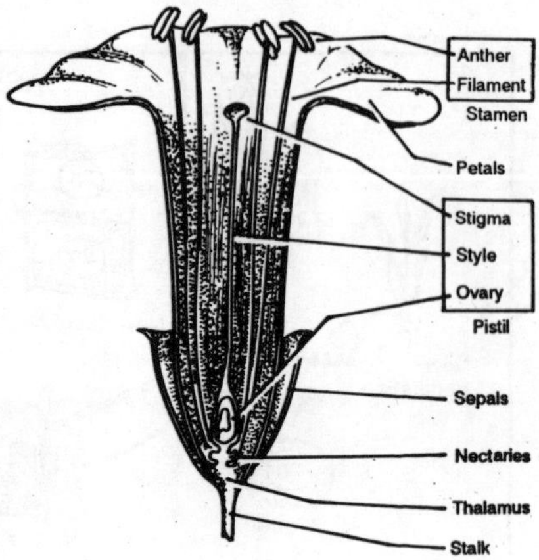

Figure 2.2 Parts of a typical flower

delicately coloured, and attractive part of the flower. From species to species the corolla exhibits an immense variety of pattern, size, and colour. Many ornamental plants are valued for their large and showy corolla. The **stamens** (collectively **androecium**) are the male reproductive structures; they consist of a thin stalk (**filament**) and **anthers** in which the **pollen grain** or haploid male gametophyte is produced. When the pollen is mature, it is discharged through small pores or when the anther walls rupture.

At the base of the corolla (or with some flowers the filaments), are glandular structures that excrete sweet, viscous fluid (**nectar**) that bees collect to produce honey. These structures are called **nectaries**. The nectaries of most flowers also synthesize aromatic substances for fragrance. The **pistil** (collectively **gynoecium**) is the female reproductive structure, and occupies the central position in the flower. A simple pistil (also called **carpel**) consists of only a single carpel, while a compound pistil contains more than one fused carpel. Depending on the species, there may be one or more pistils per flower. The pistil consists of the ovary, on which a slender elongated region or **style** is located. At the free end of the pistil there is usually a flattened tip called the **stigma**. Inside the ovary there are one or more **ovules**, which are attached by a short stalk (**chalaza**) to the ovary wall.

A flower is termed **complete** when it contains sepals, petals, stamens, and pistils; if any one or more of these four whorls is absent, it is called **incomplete**. A flower is **perfect** and **hermaphrodite** (bisexual) when it carries stamens as well as carpels, even though it may be **incomplete**, lacking the corolla. If either the androecium or gynoecium is absent, the flower is **unisexual**. If the stamens are absent or non-functional, the flower is classified as **pistillate** and **staminoid** (female flowers). Flowers lacking pistils are called **staminate** and those with non-functional pistils are called **pistilloid** (male flowers). These are **unisexual** or **imperfect** flowers.

Flowers grow on plants in many different ways; the most common is that they develop in clusters of various sizes and forms. The term **inflorescence** refers to a flower cluster. Some common inflorescence types are shown in Figure 2.3. There are species, however, in which flowers arise singly (solitary), as in cotton, soybean, or peanut. For more detailed reading on the botany of flowers, the reader is referred to Ahmad (1984).

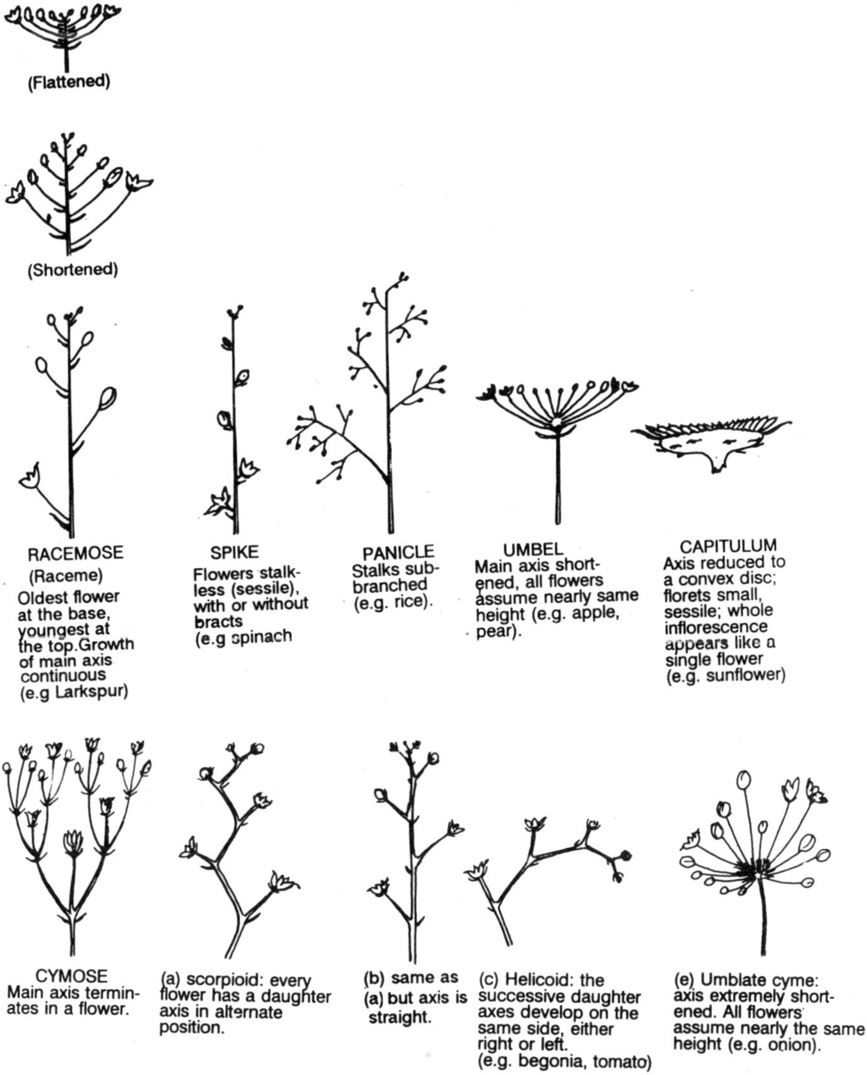

Figure 2.3 Common inflorescence types

Opening or blooming of the flower (**anthesis**) is a highly sensitive process. It is strongly influenced by many different factors, including day length, duration and intensity of illumination, ambient temperature, plant age, soil moisture, and other plant morphogenic factors (Rahman 1952).

2.3 Pollination and fertilization

We have seen that pollen and egg are located in separate structures within a flower—in the anther and ovary, respectively. In some species these floral parts may be located in the same flower, as for example in cotton and wheat flowers; whereas in some other species (**monoecious**) the carpel(s) and stamens may be located in separate flowers on the same plant, like the staminate and pistillate flowers of maize, cucurbits, oat, and walnut. They may even be found in flowers on separate plants (**dioecious**), e.g. date palm, spinach, and sugar beet. It is, therefore, necessary that the male gamete (sperm) carried in the pollen and the female gamete (egg) present in the ovule reach one another by some mechanism. As a matter of fact, the pollen becomes highly mobile. It is set free from the anthers to move about carried by air, insects, or other agents and be deposited on the stigma. The egg remains fixed inside the ovary where it waits to receive the sperm. The act of transfer of pollen from the anthers to the stigma is termed **pollination**; it is a crucial step for sexual reproduction in plants. As we shall see later, the stamens and carpels not only produce the gametes but also play a vital role in the process of pollination.

2.3.1 Floral mechanisms

The pollination of a flower is determined by its structural organization, although there is considerable overlap in behaviour even among different members of the same species. The following scheme describes the general features of flowers of common crop species with respect to their pollination behaviour.

Figure 2.4 *Helianthus annuus* (sunflower).

1. Cleistogamy. Flowers do not open and are internally self-pollinated and self-fertilized. Examples are species of *Viola*, *Oxalis*, and many grasses including some types of wheat and barley.

2. Chasmogamy. Pollination occurs after the flowers have opened. Two subtypes can be distinguished in this group.

1. **AUTOGAMY:** Flowers are hermaphrodite (bisexual) and pollen is transferred from anthers to the stigma of the same flower either directly or by some agent like insects or wind. This is called self-pollination. Some important examples are wheat, rice, and soybean.

2. **ALLOGAMY:** Cross-pollination with pollen from a different flower of the same species or occasionally the same genus. *Brassica* species, alfalfa, rye, and sunflower have flowers which are naturally cross-pollinated (Fig. 2.4). In sunflower, the whorl of large yellow petals or ray-florets are staminate. The yellow dots in the inner whorls are hermaphrodite, protandrous disc florets, a situation which favours allogamy.

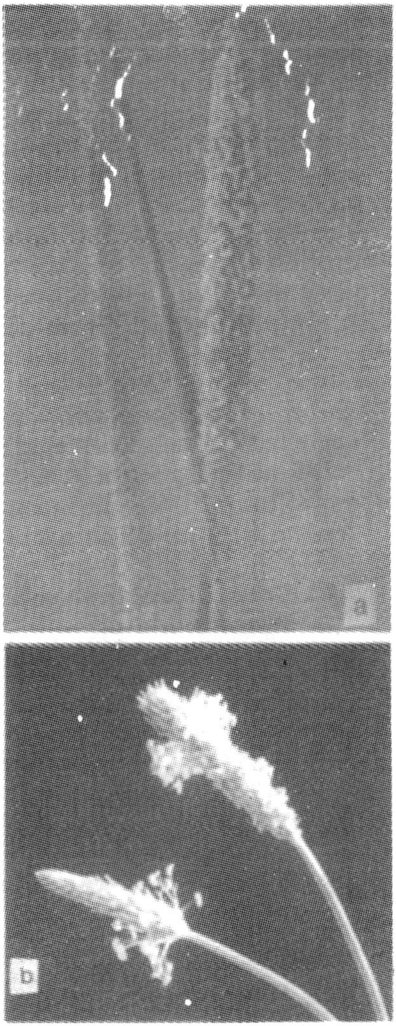

Figure 2.5 (a) Protogyny in *bajra* (*Pennisetum typhoides*); (b) Protogyny in *Plantago lanceolata*. (Photographs by author.)

Cross-pollination in most crop plants usually occurs by means of wind (**anemophily**) or insects (**entomophily**). A large variety of insects including bees and wasps (Hymenoptera), butterflies and moths (Lepidoptera), flies (Diptera), beetles (Coleoptera), bugs (Hemiptera), and even some birds are attracted to flowers by their colour, fragrance, nectar, and edible pollen and sap. In a field or crop canopy these creatures visit a large number of flowers and the pollen sticks to their body parts and is thus transferred to the stigma of companion flowers. Some reasons for allogamy are:

a. *Self-incompatibility.* The pollen coming from the plant's own flowers fails to germinate because of physiological or genetic reasons, or else the stigma and anthers are isolated in such a way that autogamy becomes mechanically impossible (e.g. many orchids).

b. *Dichogamy.* The ripening times of pistil and anthers in a flower or inflorescence do not coincide. There are species in which the anthers ripen before the carpels, as in pearl millet (*bajra*); this situation is called **protandry**. In other instances, the stigmas mature earlier than the stamens; the term **protogyny** refers to this situation. A characteristic example of protogyny is *Plantago*. The stigmas in a section of the inflorescence emerge and mature earlier than the anthers. In Figure 2.5a, on the left-hand ear only stigmas have emerged. The right-hand ear is full of anthers but the stigmas have withered off long before. Because of such unsynchronized maturity times of stigma and anthers, self-pollination is not possible.

c. *Cytogenetic reasons.* There are a number of cytogenetic factors which affect the formation and functioning of gametes and therefore also the pollination pattern. Chromosomes involving translocations, inversions, aneuploidy, and autopolyploidy directly affect the fertility of the pollen or the egg or both. Plant breeders must thoroughly understand these features. A brief discussion is given here, but for detailed information the student is referred to Elliott (1958) and Swanson et al. (1981).

Various species and their pollination behaviour are listed in Table 2.1.

Table 2.1 Selected species with pollination behaviour and chromosome number

English name	Botanical name	Flower type	Chromosome number (2n)	Natural pollination type
		GRASSES		
Barley	*Hordeum vulgare*	⚥	14	10% cross
Corn (maize)	*Zea mays*	♂♀	20	Cross (monoecious)
Millet (pearl)	*Pennisetum typhoides*	⚥	14	Cross (protogynous)
Millet (finger)	*Eleusine coracana*	⚥	36(72)	Self
Millet (Proso)	*Panicum miliaceum*	⚥	36(72)	–
Millet (foxtail)	*Setaria italica*	⚥	18	Self
Oats	*Avena sativa*	⚥	18	Self
Rice	*Oryza sativa*	⚥	24	>1% cross
Rice (wild)	*Oryza fatua*	⚥	48	>1% cross
Rye	*Secale cereale*	⚥	14	Cross
Sorghum	*Sorghum bicolor*	⚥	20	10% cross
Wheat	*Triticum aestivum*	⚥	42	<5% cross
Wheat (macaroni)	*Triticum turgidum*	⚥	28	<5% cross

English name	Botanical name	Flower type	Chromosome number (2n)	Natural pollination type
Triticale	*Triticosecale* spp.	⚥	42, 56	Self
Sudan grass	*Sorghum bicolor sudanensis*	⚥	20, 40	>25% cross
Napier grass	*Pennisetum purpureum*	⚥	28	Cross
	LEGUMES			
Beans	*Phaseolus vulgaris*	⚥	22	Self
Broad beans	*Vicia fabia*	⚥	12, 14	Self
Chick pea or gram	*Cicer arietinum*	⚥	16	Self
Cowpea	*Vigna sinensis*	⚥	22, 24	Self
Lima beans	*Phaseolus limensis*	⚥	22	Self
Mung beans (*mash*)	*Vigna mungo*	⚥	22	Self
Mung beans (*mung*)	*Vigna radiata*	⚥	22	Self
Pea	*Pisum sativum*	⚥	14	Self
Peanut (groundnut)	*Arachis hypogaea*	⚥	20	Self
Soybean	*Glycine max*	⚥	80	Self
Lentil	*Lens culinaris*	⚥	14	Self
Pigeon pea	*Cajanus cajan* (= *C. indicus*)	⚥	22, 44, 66	5–40% cross
	FORAGE CROPS			
Alfalfa (lucerne)	*Medicago sativa*	⚥	32	Cross
Alfalfa (variegated)	*Medicago media*	⚥	32	Cross
Alfalfa (yellow)	*Medicago falcata*	⚥	32, 16	Cross
Clover (*berseem*)	*Trifolium alexandrinum*	⚥	16	Self
Clover (red)	*Trifolium pratense*	⚥	14	Cross
Clover (white)	*Trifolium repens*	⚥	32, 48	Cross
Clover (crimson)	*Trifolium incarnatum*	⚥	14	Cross
Clover (sweet)	*Melilotus officinalis*	⚥	16	Cross
Clover (bush)	*Lespedeza striata*	⚥	18	–
Clover (Persian)	*Trifolium resupinatum*	⚥	16	–
Guar (guara)	*Cyamopsis tetragonoloba*	⚥	14	–
Bermuda grass	*Cynodon dactylon*	⚥	18, 36	Cross
Brome grass	*Bromus inermis*	⚥	56	Cross
Blue grass	*Poa pratensis*	⚥	42	Cross
Birdsfoot trefoil	*Lotus corniculatus*	⚥	12, 24	Cross
Timothy	*Phleum pratense*	⚥	42	Cross

English name	Botanical name	Flower type	Chromosome number (2n)	Natural pollination type
Bahia grass	*Paspalum notatum*	⚥	40	Cross
Orchard grass	*Dactylis glomerata*	⚥	28	Cross
Rye grass (perennial)	*Lolium perenne*	⚥	14	Self-incompatible
		FIBRE CROPS		
Cotton (Asiatic)	*Gossypium arboreum*	⚥	26	Self
Cotton (desi)	*Gossypium herbaceum*	⚥	26	5–25% cross
Cotton (American–Pima)	*Gossypium barbadense*	⚥	52	5–25% cross
Cotton (American–upland)	*Gossypium hirsutum*	⚥	52	5–25% cross
Hemp (marijuana)	*Cannabis sativa*	⚥	20	Cross
Jute	*Corchorus capsularis*	⚥	14	Self
Jute	*Corchorus olitorius*	⚥	14	Self
		OILSEED CROPS		
Flax (linseed)	*Linum usitatissimum*	⚥	30	Self
Toria	*Brassica campestris*	⚥	20	Cross (self-sterile)
Rape (yellow) (*sarson*)	*Brassica campestris*	⚥	20	Self
Summer rape (*gobhi sarson*)	*Brassica napus*	⚥	38	Self
Brown mustard (Indian *raya* or *rai*)	*Brassica juncea*	⚥	36	Self
Taramira (Jemba, rocket)	*Eruca sativa*	⚥	22	Cross (self-sterile)
Abyssinian mustard	*Brassica carinata*	⚥	34	Self
Sunflower	*Helianthus annuus*	⚥	34	Cross
Safflower	*Carthamus tinctorius*	⚥	24	Cross
Soybean	*Glycine max*	⚥	40	Self
Groundnut (Peanut, *moongphali*)	*Arachis hypogaea*	⚥	40	Self
Coconut	*Cocos nucifera*	♂, ♀	32	Cross (dioecious)
Oil palm	*Elaeis guineensis*	♂, ♀	32	Dioecious
		VEGETABLE CROPS		
Cabbage	*Brassica oleracea*	⚥	18	Cross (incompatible)

English name	Botanical name	Flower type	Chromosome number (2n)	Natural pollination type
Carrot	*Daucus carota*	⚥	9	Dioecious
Cauliflower	*Brassica oleracea botrytis*	⚥	18	Cross
Celery	*Apium graveolens*	⚥	22	30% cross
Cucumber	*Cucumis sativus*	⚥	14(28)	Monoecious
Radish	*Raphanus sativus*	⚥	18	Cross (incompatible)
Turnip	*Brassica rapa*		18	Cross
Eggplant	*Solanum melongena*	⚥	24	Self
Lettuce	*Lactuca sativa*	⚥	18	Self
Okra	*Hibiscus abelmoschus/esculentus*	⚥	36	Often
Pepper	*Capsicum frutescens*	⚥	19–20	5% cross
Tomato	*Lycopersicon esculentum*	⚥	24	Self
Spinach	*Spinacia oleracea*	♂, ♀	12(24)	Cross (dioecious)
Onion	*Allium cepa*	⚥	16	Self
Potato	*Solanum tuberosum*	⚥	48	Self (apomixis common)
Garlic	*Allium sativum*	⚥	16	Self
Zingiber	*Zingiber officinale*	⚥	22	Self
	FRUIT CROPS			
Guava	*Psidium guajava*	⚥	22	Self
Orange	*Citrus sinensis*	⚥	18	Cross
Peach	*Prunus persica*	⚥	16	Cross
Apricot	*Prunus armeniaca*	⚥	16	Cross
Plum	*Prunus americana*	⚥	16	Cross
	Prunus domestica	⚥	48	Cross
Apple	*Malus sylvestris* (= *M. pumila*)	⚥	34, 51	Cross
Banana	*Musa acuminata*	⚥	22, 33, 44	Parthenogenetic
Banana	*Musa paradisiaca* (= *M. sapientium*)	⚥	22, 33, 44	Parthenocarpic

English name	Botanical name	Flower type	Chromosome number (2n)	Natural pollination type
Date palm	*Phoenix dactylifera*	♂, ♀	36	Cross (dioecious)
Fig	*Ficus carica*	⚥ + ♂, ♀	26	Cross (parthenogenetic)
Grape	*Vitis vinifera*	⚥	38	Self
Olive	*Olea europea*	⚥	46	–
Mandarin	*Citrus reticulata*	⚥	18	Cross (apomixis)
Pear	*Pyrus communis*	⚥	34	Cross
Papaya	*Carica papaya*	♂, ♀	18, 36	Cross (dioecious)
Strawberry	*Fragaria ananassa*	⚥	56	Cross (rarely polygamous)
Watermelon	*Citrullus vulgaris*	⚥, ♂+♀	22	Cross (monoecious)
Mango	*Mangifera indica*	♂+⚥	40	Cross
Loquat	*Eriobotrya japonica*	⚥	34	Cross (polygamous)
		SUGAR CROPS		
Sugar beet	*Beta vulgaris*	♂, ♀	18	Cross (dioecious)
Sugarcane	*Saccharum officinarum*	⚥	80	Cross

2.3.2 Genetic factors influencing pollination behaviour

1. **Polyploidy.** A diploid species, in its somatic cells, generally has chromosomes of each kind in pairs. A tomato plant, for example, has 2n=2x=24 chromosomes representing 12 pairs, one pair each of chromosomes Nos. 1–12. The members of these pairs are called homologous chromosomes. In this species, therefore, the set of 12 chromosomes is the **haploid,** and presumably, every individual chromosome in this set should be structurally and genetically different from every other one (Fig. 2.6–a, b, c). Haploid sets are actually carried in the male and female gametes, which on fertilization produce a **diploid**. Several situations are now known in which the chromosome number of a certain species can be multiples of the basic haploid number. The barley species *Hordeum bulbosum*, for instance, has 2n=28 (4 × 7, the basic haploid number); such a species is called **tetraploid**. Assuming a self-fertilizing species with genomic sets AA, doubling would lead to AAAA or an **autotetraploid**. A hybrid AB from two different species, say AA × BB, will after doubling be an **allotetraploid** AABB. Many plant

genera, including cultivated species, have polyploid chromosome numbers. The *Triticum* genus includes species with the somatic numbers 14, 28, 42, and 56. Similarly the rose genus (*Roza*) has species with 14, 21, 28, 35, 42, and 56 chromosomes, giving diploid (2n=2x=14), triploid (2n=3x=21), tetraploid (2n=4x=28), pentaploid (2n=5x=35), etc.

Polyploid varieties with an even number of genomes, particularly the allopolyploids, are often fully fertile. This is because during meiosis the synapsis between different homologous chromosomes remains normal, and balanced fertile gametes are produced. Triploids are commonly created by crossing a normal diploid, whose gametes are n=x, with a tetraploid, the gametes of which are 2n=2x. The triploid plants fail to produce normal gametes because of severe synaptic irregularities during meiosis. Thus even though the other floral mechanisms may be normal, these plants are highly sterile. In such cases, therefore, the crop is either grown from seed continuously produced by hybridizing 2n × 4n types, as in the case of seedless watermelons, or propagated by vegetative methods, as with the hybrid of *bajra* × Napier grass (Khan and Rahman 1963) (Fig. 2.7). In Figure 2.7, plate (a) shows anaphase I in *bajra*, with seven I's (monovalents) on each side; in plate (b) 14 II's (bivalents) in metaphase I of Napier grass are visible; plate (c) reveals complete failure of pairing in the triploid hybrid; and plate (d) shows seven II's and seven I's in the hybrid.

2. Aneuploidy. The chromosome number of a species may change in small decrements or increments involving one or two chromosomes. The designation 2n is used to indicate a **disomic** chromosome number, and 2n+1 and 2n−1 to indicate, respectively, **trisomic** and **monosomic** plants. For example, if in a disomic (2n) barley plant one chromosome 3 is missing, it is termed monosomic–3. Similarly, if the monosomic–3 plant carries a third additional chromosome 5, it will be designated as monosomic–3, trisomic–5. If both members of a certain homologous pair are missing, it becomes nullisomic, i.e. 2n−2. Experiments from a wide variety of crops have shown that the viability of n−1 gametes and 2n−1 zygotes and even the seed is considerably reduced.

A thorough discussion of the origins, cytogenetics, and applications of plant breeding has been given by Khush (1973).

3. Chromosome rearrangements (aberrations). The linear arrangement of genes in a chromosome remains constant for every individual in a species. Very infrequently under natural conditions, or on exposure to strong ionizing radiation such as X–rays or gamma rays, the chromosomes are fractured, and if the breaks are few the fragments may rejoin in several aberrant manners. Figure 2.8 shows some types of chromosome aberrations: (a) represents an assumed standard segmental sequence of two different homologous chromosome pairs exposed to ionizing radiation. Radiation causes random breaks (represented as gaps), and the broken sections may rejoin to

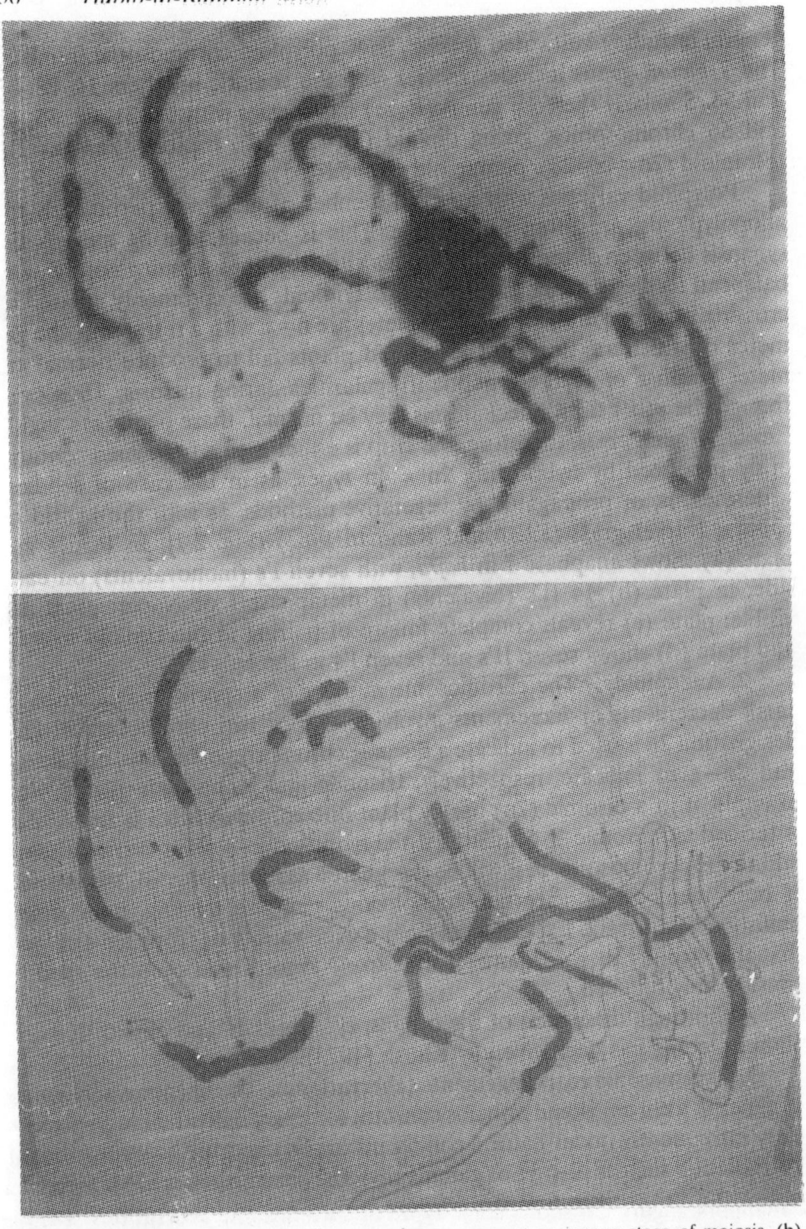

Figure 2.6 (a) Photomicrograph of tomato chromosomes at pachytene stage of meiosis. (b) Schematic diagram of (a), identifying the 12 chromosome pairs. (Reproduced from Rick (1978) by permission of the author.)

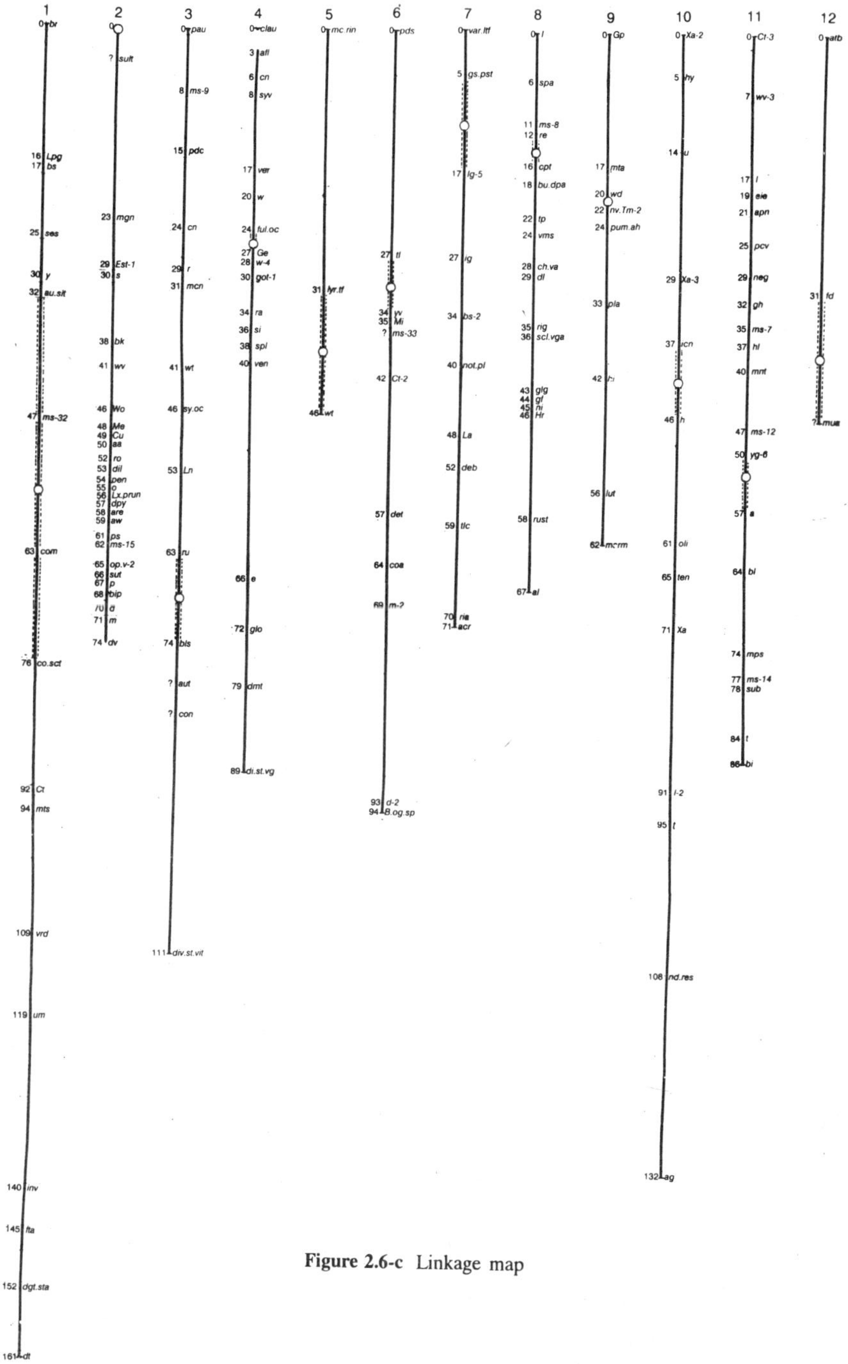

Figure 2.6-c Linkage map

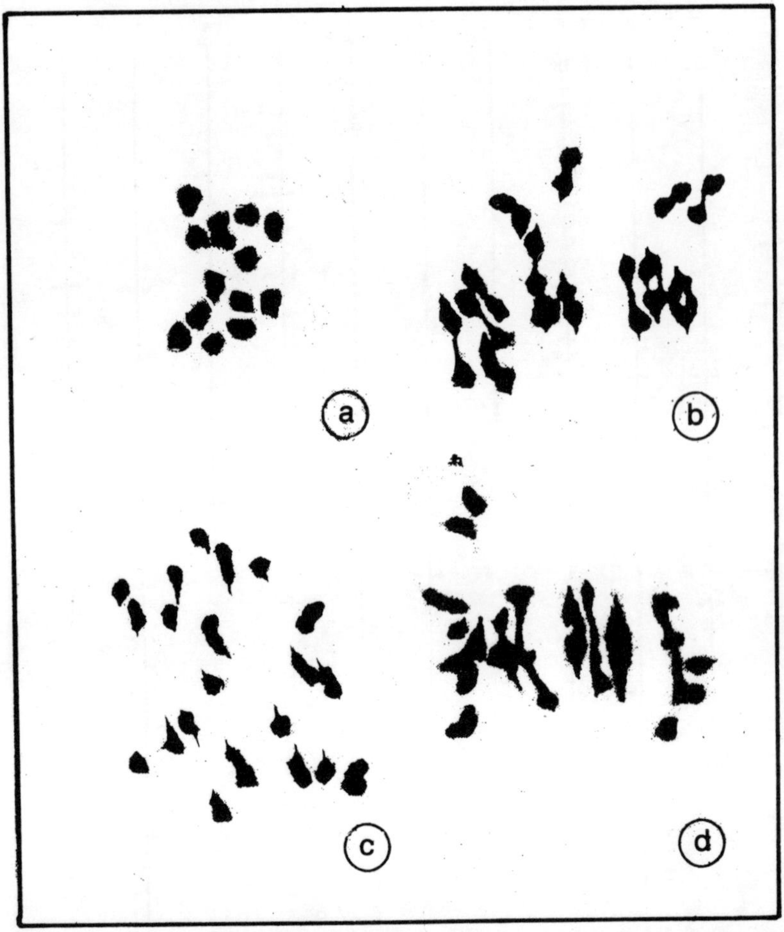

Figure 2.7 Chromosomes of a triploid hybrid derived by crossing *bajra* (*Pennisetum typhoides*) × Napier grass (*P. purpureum*) (Khan, M. D. and Rahman, H. 1963).

produce aberrant or rearranged segmental sequences. (b) **Terminal deletion** where the end segments AB are completely lost. **Interstitial deletion** arises if segment CD is eliminated and the terminal portion rejoins with the main section. (c) A **duplication** may occur if the broken section <u>CD</u> is inserted in adjacent tandem or adjacent inverse sequence. (d) **Inversion** arises from breaks on two points along the chromosome length followed by rejoining of the broken pieces in inverted sequence. An inversion is *paracentric* if the centromere is excluded and *pericentric* if the centromere is included in the inverted piece. (e) A segment from one chromosome is **translocated** into a non-homologous chromosome.

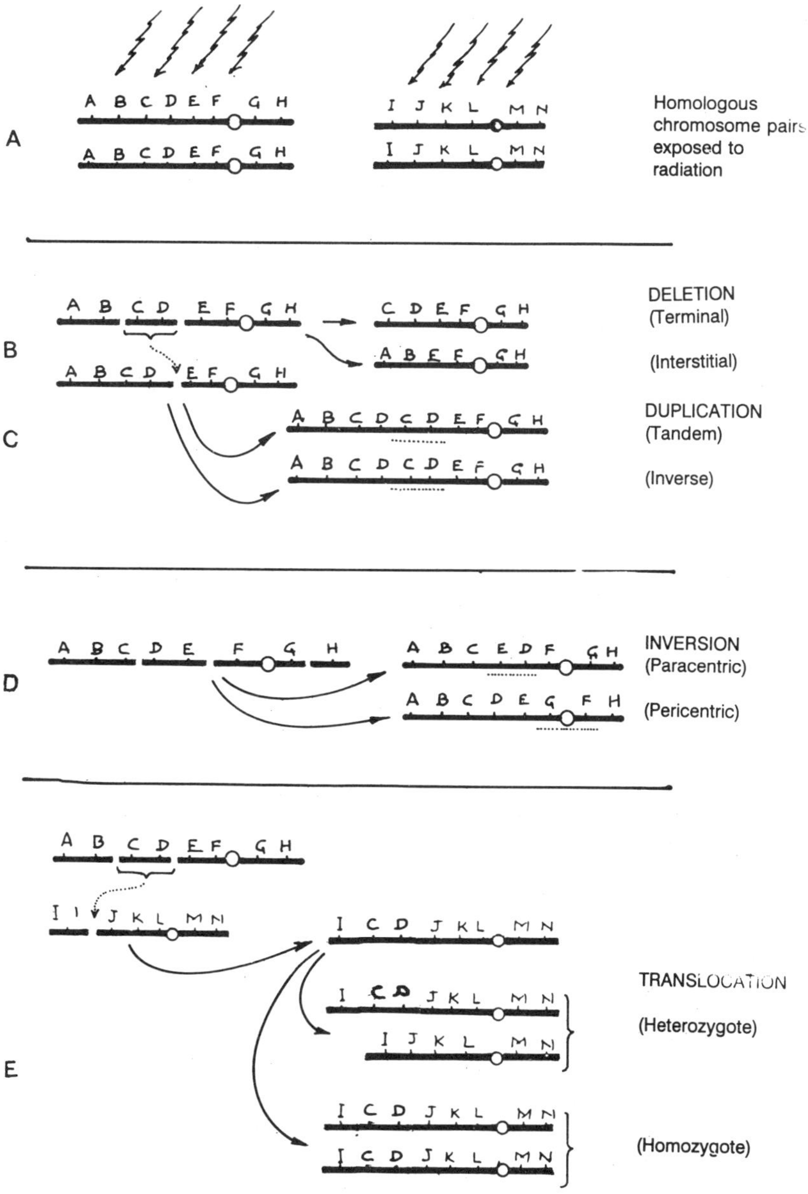

Figure 2.8 Various chromosomal aberrations

Cytological examination of such cells reveals deficiencies (deletions), duplications, translocations, and inversions. Each of these conditions implies peculiar genetic and cytogenetic consequences (for details see Swanson et start here al. 1981). There are, however, common principles which determine the stability or survival of chromosome aberrations. Each rearranged chromosome must contain only one centromere. Chromosome fragments devoid of a centromere are eliminated during cell division. Similarly, new chromosomes formed by the union of fragments are eventually lost if they include two centromeres. Cytological examination of dividing cells exposed to radiation often shows chromosome fragments which are excluded from the nuclei because they lack a centromere or have more than one centromere. If more than one centromere is present, deletions of blocks of genes can develop. Such a dicentric (having two centromeres) chromosome forms a bridge across the metaphase plate with each centromere pulling to the opposite pole. During early telophase the bridge breaks at unspecified points; thus each daughter nucleus receives unequal lengths of the chromatid involved. This type of bridge–breakage–reunion cycle continues in subsequent generations, resulting in deletion, duplication, and acentric fragment formation, culminating in the loss of whole chromosomes. These are major chromosomal imbalances and cause the cell to die. Particularly on the pollen side, even small chromosomal deficiencies in pollen prove lethal. A decrease in the fertility of pollen, egg, or even zygote can, therefore, sometimes be attributed to chromosome aberrations.

Direct effects of chromosome aberrations on the reproductive system. Each type of chromosome aberration produces different effects. The deletion of chromosomes characteristically has the following results. (i) Crossing-over is suppressed, particularly in the region spanning the deficiency. (ii) A deleted chromosome can never revert to its original condition. (iii) A recessive gene located on a normal chromosome is expressed when its allele is missing because of a deletion in the same region from the homologue, a situation called **pseudo-dominance**. (iv) Deletion loops are produced during chromosome synapsis in meiosis.

The different types of inversions produce somewhat different effects. Inversion heterozygotes reduce genetic recombination by inhibiting chromosome synapsis during meiosis and also by selectively eliminating the products of crossovers occurring in the inverted regions. Gametes with translocations normally abort, causing a situation of semi-sterility which can be identified from the presence of a significant number of shriveled, abnormal pollen grains. Deletions, inversions, and translocations, therefore, generate semi-sterility by inducing gross irregularities in meiosis and the production of imbalanced, highly mortal gametes. These conditions may even give rise to lethal zygotes.

2.3.3 Flower structures promoting self-pollination

The organization of the various floral parts and the time of anthesis strongly influence the mechanism of pollination. Most plants show cleistogamy, in which pollination takes place before the opening of flowers, thus assuring self-pollination. Several cultivated grasses manifest this behaviour; a common example is brome grass (*Bromus carinatus*). Cereal crops including varieties of wheat, oat, and barley sometimes exhibit cleistogamy, and pollination may have already occurred before the spikes actually emerge from the boot. Cleistogamous flowers also occur in lespedeza (*Lespedeza striata*), some types of violets, and certain weeds, e.g. some silene species.

The position of the anthers in relation to the stigma is another feature that favours self-pollination. Cotton and other members of this family are a typical example. Figure 2.9 shows a cotton flower with several whorls. The pistil occupies the superior central position with the stigma emerging out of a staminal tube. Numerous stamens develop all along the staminal tube, which is located so that pollen is shed directly on the stigma. The flowers have a large number of stamens, and filaments of all the anthers are joined to form a tube or staminal sheath through which the young pistil grows out exposing the stigma for pollination. The pollen grains from the anthers fall directly on the receptive stigma with even a slight jerk. In the field, cotton flowers are frequently visited by large numbers of insects; thus a good proportion of flowers get cross-pollinated, even though the floral morphology favours self-pollination. For this reason cotton is classed as an often cross-pollinated crop.

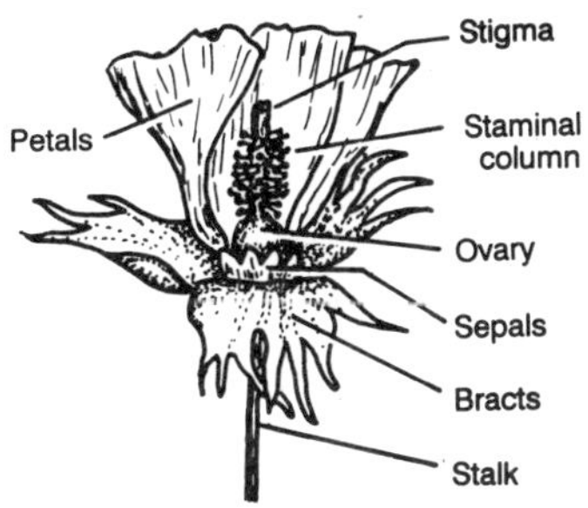

Figure 2.9 Cotton flower

2.3.4 Floral morphology supporting cross-pollination

Most cultivated species are naturally adapted to cross-pollination. Several conditions exist in plants that encourage cross-pollination. Some of the most common situations are described here.

1. Dioecious plants. When male and female flowers are located on separate plants, the plants are called **dioecious**. This means that some plants are male because they bear only staminate flowers, while others are female since only pistillate flowers are borne on them. This always results in cross-

Figure 2.10 The staminate (left) and pistillate (right) inflorescences of maize (photographs by author).

pollination between the members of a plant population. Some examples of this type are spinach, sugar beet, hemp, hops, asparagus, poplar, willow, and date palm.

2. Monoecious plants. Plants in which male (staminate), and female (pistillate) flowers are separate but located on the same plant are called **monoecious**. This situation is found in maize, castor bean, and vine crops of Cucurbitaceae. The maize plant, for example, bears staminate flowers in the tassel and pistillate flowers in the cob (Fig. 2.10). Wind and insects effect cross-pollination up to more than 90 percent of the time.

3. Self-pollination prevented by floral structure.

A. DICHOGAMY. In some species the stamens and pistils of flowers in an inflorescence mature at different times. One condition is **protandry**, in which the anthers within a flower mature earlier than the stigma. The pollen shed by these anthers, therefore, pollinates flowers on other branches or other plants but not their own. Many plant species show **protogyny**, in which the stigma becomes receptive much sooner than the anthers of the same plant can release pollen. The maturity difference between the anthers and pistils is usually two or three days. This time is long enough for cross-pollination to effectively occur. These situations are observed in numerous plant species including many crops like maize (*Zea mays*), *bajra* (*Pennisetum typhoides*), Napier grass (*Pennisetum purpureum*), lespedeza (*Lespedeza stipulacea*), guayule (*Parthenium argentatum*), and plantain (*Plantago lanceolata*).

B. HETEROSTYLY. This is a floral modification in which the stamens and pistils are of unequal lengths. This variation may occur among different plants of the same species, or within a single plant. Some flowers contain stamens with very short filaments and long styles, while others have flowers with long filaments but very short styles. Two types of flowers are common (Fig. 2.11). **Pin flowers** have long pistils and short stamens. **Thrum flowers** have short pistils and long stamens. A range of intermediate types like long, medium, or short styles; or long, medium, and short stamens also exist in various genera. Some well known examples of heterostyly are linseed (*Linum usitatissimum*), *Polygonum, Oxalis, Primula,* thyrum, and buckwheat (*Fagopyrum vulgare*).

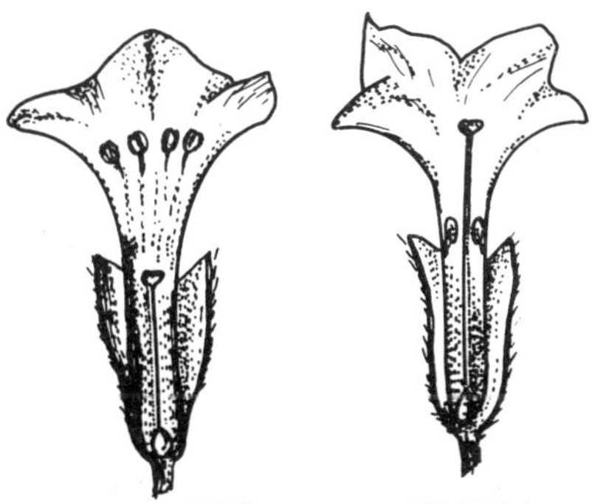

Figure 2.11 Pin flower and thrum flower

Still another kind of modification is observed in Natal plum (*Carissa grandiflora*). In this species, plants with short styles have normal anthers that produce functional pollen, but plants with long styles have poorly developed anthers devoid of pollen. These plants which produce no pollen are thus male sterile, and 100 percent cross-pollination is assured.

2.3.5 Incompatibility mechanisms

The normal mode of pollination may be altered because of various genetic and cytoplasmic mechanisms. These mechanisms are quite common in crop plants, and have been reported in tobacco (*Nicotiana*), tomato (*Lycopersicon*), clovers (*Trifolium* spp.), sugar beet (*Beta vulgaris*), sunflower (*Helianthus annuus*), *Brassica* spp., onion (*Allium* spp.), and many species of forage grasses. The breeder should be aware of how these processes operate, as they will have direct effect on the methods of pollination control and breeding procedures that can be adopted.

Incompatibility may be such that pollen becomes incapable of fertilizing the eggs within its own flowers, although both appear structurally normal. This is called **self-incompatibility**, and it limits self-pollination or otherwise promotes cross-pollination. Conversely, **cross-incompatibility** is between the pollen and gynoecium of different plants, and prevents cross-pollination. Incompatibility is caused by some physiological process which brings about a restriction in the normal process of fertilization. This can happen in two ways: i) Pollen on the stigma fails to germinate. ii) Pollen may germinate but tube growth down the stigma is so slow that it does not succeed in reaching the ovule at the proper time, so the egg loses viability and no fertilization occurs.

Incompatibility has been discovered in many families of cultivated plants such as Rosaceae, Liliaceae, Leguminoseae, Solanaceae, and Graminae. It is genetically controlled, and the common type called **personate incompatibility** (e.g. in *Nicotiana*) shows monogenic inheritance. The gene controlling it has been assigned the symbol S, and it shows multiple allelomorphism. This gene has a large number of allelomorphs designated S_1, S_2, S_3, etc., or S_n, S_m, etc. A diploid plant can carry any one pair of these alleles, whereas a haploid pollen will contain only one of them. When identical alleles are present in both the pollen and the pistil, the two become incompatible. Pollen germination or tube growth through the style is inhibited, causing failure of fertilization. For pollination to occur, therefore, the S allelomorph in the pollen tissue must be different from that in the stylar tissue.

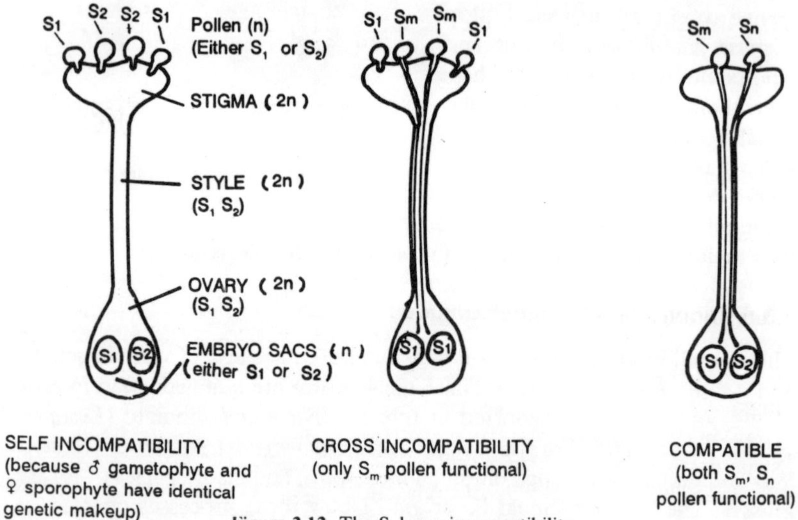

Figure 2.12 The S–locus incompatibility

Assume that the plants to which pistils A, B, and C (Fig. 2.12) belong have the genotype S_1S_2. (a) In this case, the pollen and eggs have either S_1 or S_2 alleles. When self-pollination occurs, the pollen fails to complete fertilization because the stylar tissue (being diploid) contains both genes identical to those of the pollen. (b) In this situation, S_m pollen germinates and ensures fertilization, but S_1 pollen coming from other plants is inhibited, hence the zygotes can be either S_1S_m or S_2S_m. This shows cross-incompatibility. (c) In this case, either S_m or S_n pollen germinates freely because neither of them has identical genes in the stylar tissue. The possible zygotes in this case will be S_1S_m, S_1S_n, S_2S_m, and S_2S_n.

In many genera of cultivated plants, a multiple allele type of incompatibility has been observed that causes considerable confusion to breeders, especially when wide crosses or seed production in F_1 or the segregating progenies in advanced generations are involved. For a more detailed account of this subject the reader is referred to Frankel and Galun (1977).

2.4 Male sterility

2.4.1 Genetic and cytoplasmic male sterility

Male sterility is caused by either genetic or cytoplasmic factors. In certain plant species, both of these mechanisms coexist. Their apparent effects are: (i) Anthers as a whole, along with the sporogenous tissue, do not develop properly. (ii) Anthers and pollen develop normally, but the anthers fail to dehisce. (iii) Anthers are apparently normal, but they contain undeveloped sporogenous tissue and consequently non-functional pollen. These factors detrimentally influence the growth and development of anther-related tissues and also anther functions.

In Figure 2.13, photomicrographs of longitudinal sections of male-fertile (a) and male-sterile (b) anthers of barley msg 16 line are presented. In (a) the developing pollen grains which are at the uninucleate stage appear large and spherical, with a distinct exine and intine; and they are also in close contact with the tapetal tissue. In (b) the pollen grains are deformed, the exine and intine are rudimentary, and they have lost contact with the tapetum. Whatever the reasons for the condition, male-sterile flowers utilize pollen exclusively from other sources. This set of factors results in outbreeding.

Male sterility mechanisms override many of the already discussed structural features of flowers that regulate pollination behaviour. Male sterility is exploited extensively in breeding programmes, particularly in mass-scale hybridization or commercial production of hybrid varieties. This has been done with such crops as maize, sorghum, onion, and tomato. Virtually all species of crop plants have been shown to produce some male-sterile variants. In addition to the possibility of exploiting them for plant breeding, male-sterile genes are of considerable value in genetic studies, for they can

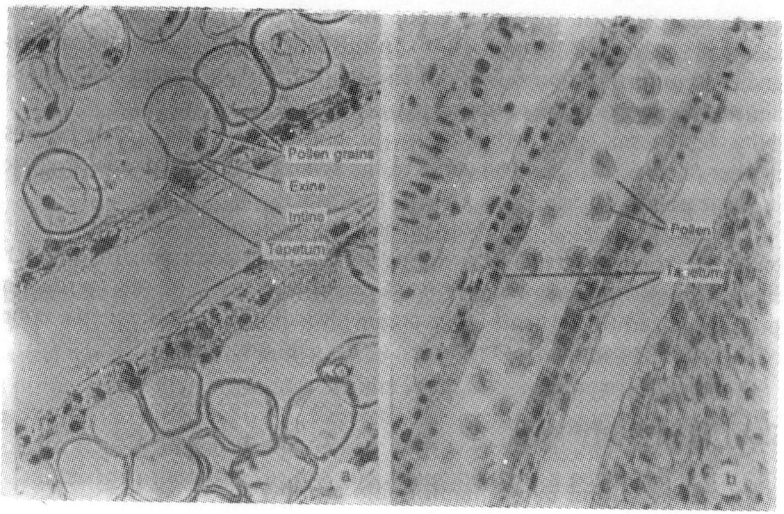

Figure 2.13 Longitudinal sections of (a) male-fertile and (b) male-sterile barley anthers.

be used to provide an excellent model for investigating the functions of cell organelles and other cell components in development.

Three different inheritance systems governing the male-sterile (*ms*) character have been identified (Jain 1959). Genetic male sterility is entirely under the control of rare, recessive, nuclear genes; whereas cytoplasmic male sterility is under the control of cytoplasmic genes and follows a maternal inheritance pattern. The third type requires the interaction of both nuclear and cytoplasmic genes; this means that *ms ms* nuclear genotypes in male-sterile cytoplasm are required for its expression.

Genetic male sterility has been found in several cultivated crops including barley, rice, sorghum, maize, wheat, squash, carrot, melon, and brinjal. Cytoplasmic male sterility is relatively less common and is rather difficult to handle in breeding programs. It has been observed in maize, sorghum, onion, sugar beet, cotton, tobacco, flax, field bean, and many other horticultural plants.

The mode of inheritance of genetic male sterility is illustrated in Figure 2.14. The genes designated as *ms* cause male sterility when present in homozygous recessive form, i.e. *ms ms* (as in variety K), while the homozygous dominant *Ms Ms* remain male-fertile (variety M). The heterozygous F_1 genotypes have normal fertility. Selfing of F_1's will produce a proportion of progeny with *ms ms* (homozygous recessive) genotypes. Such plants will be male-sterile. Genetic male sterile lines cannot be propagated alone; therefore this characteristic can be maintained by out-crossing with fertile

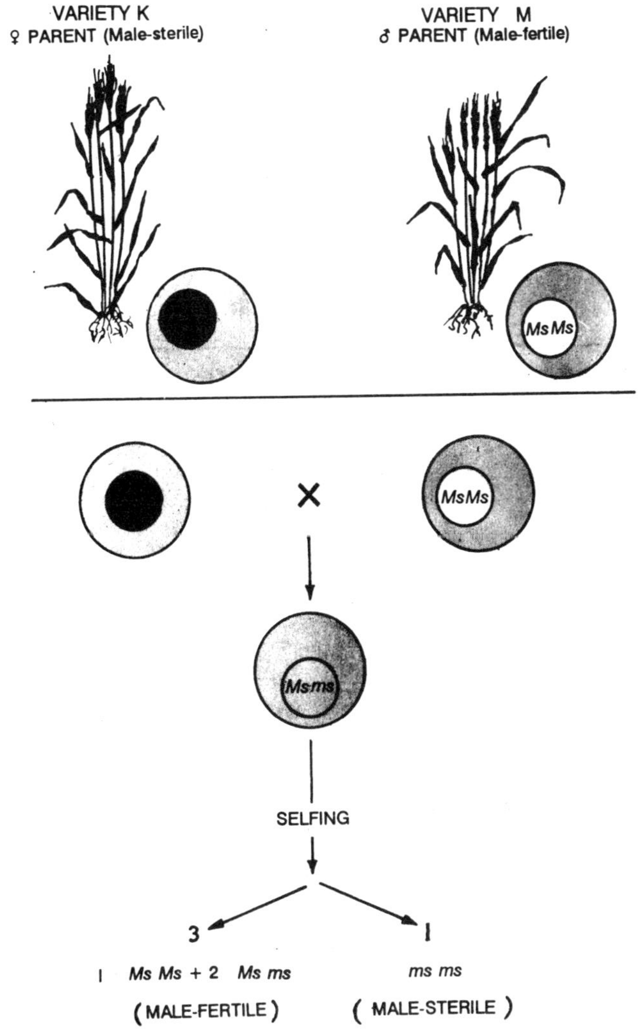

Figure 2.14 Genetic male sterility. Homozygous recessive alleles (*ms ms*) cause male sterility (variety K). Heterozygous (*Ms ms*) or homozygous dominant (*Ms Ms*) produce fertile pollen.

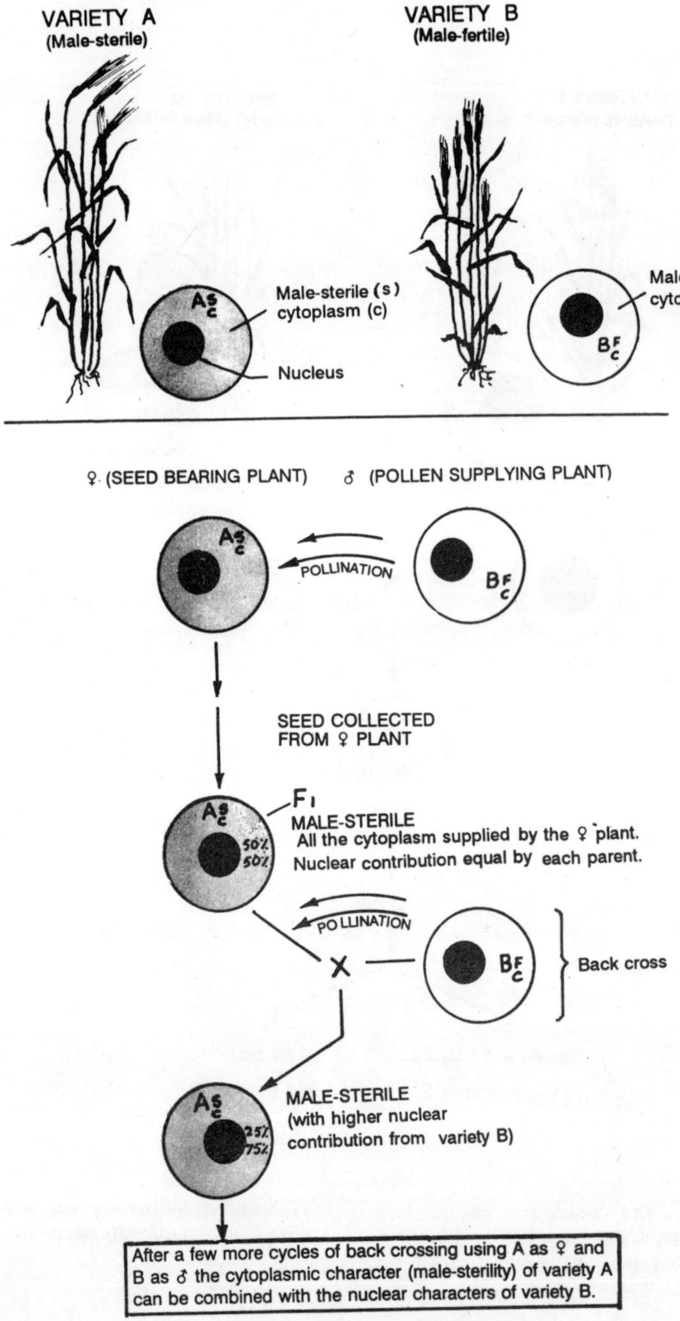

Figure 2.15 Cytoplasmic male sterility

(preferably heterozygous *Ms ms*) plants using male-sterile plants as the female parent.

In inheritance of cytoplasmic male sterility, during sexual reproduction the cytoplasm is almost completely contributed by the seed-bearing (♀) plant. If a sterility character is present in the egg cytoplasm, the progeny will be male-sterile even though the pollen comes from a fertile plant. Figure 2.15 illustrates cytoplasmic male sterility. Variety A carries cytoplasm with male-sterility factors; these plants produce male-sterile flowers. At the time of fertilization the female gamete (egg) contributes a haploid nucleus (An) and large quantity of cytoplasm (Asc); while the male parent forms (variety B), which bears male-fertile flowers, provides a male gamete through pollen, and contributes a haploid nucleus (Bn) but negligible or no cytoplasm (Bfc). This results in the female being the only source of cytoplasm for the zygote. Hence all the genetic information carried by the cytoplasm affecting developmental aspects of anthers or pollen is exclusively contributed by the female parent. Thus the progeny also exhibit male sterility irrespective of the fact that half the nuclear contribution is from the male-fertile pollen parent.

In sorghum and onion, male sterility is utilized by maintaining two populations of a certain variety. One carries a genetic factor, and the other a cytoplasmic factor for male sterility, but they can be multiplied by inbreeding through partial seed set. The hybrid from these lines is completely male-sterile, and is then used as the female for commercial hybrid production.

In barley, more than 24 different genetic male-sterile mutants have been described (Hans et al. 1971) and investigated in enough detail to indicate the nature of the morphological defects which are involved in male-sterile anther development (Frankel and Galun 1977; Mian et al. 1974). Genetic studies indicate that each of the sterility mutants exhibits markedly different morphogenetic effects, and they show monogenic recessive inheritance. Therefore, each specific gene is designated as *msg* 1 to *msg* 24. Studies by Roath and Hockett (1971) indicate that both meiosis and the formation of tetrads are normal. Differences in morphogenic details of the tapetal and sporogenous tissue first begin to appear at the onset of the free microspore stage and culminate in non-functional pollen grains.

Elaborate study by Mian et al. (1974) on successive stages of barley anther tissue development with respect to cell differentiation, meiosis, and variations in DNA and RNA content in nuclei for *msg* 5, *msg* 9, *msg* 10, *msg* 11, and *msg* 18, has shown that each gene for these endomorphic traits manifests striking differences (Fig. 2.13). Further work on *msg* 16 (Mian et al. 1981) related to a comparison of development of normal and male-sterile anthers by electron microscopy provide evidence that the male-sterile mutant fails to synthesize certain crucial metabolic substances which regulate differentiation of growing pollen grains.

2.3.2 Male gametocides

Because male sterility has profound potential uses in production of mass-scale hybrids or in promoting cross-pollination in naturally self-pollinated crops, which sometimes becomes necessary in order to maintain or enhance genetic heterozygosity, many efforts have been made to induce it by artificial means. Considerable success has been achieved through chemicals which selectively retard or inhibit the development of anthers. Ethrel (2–chloro-ethylphosphonic acid) has been found to induce male sterility in wheat, rice, barley, and other cereals. Applied before anthesis, this chemical stops anther development. For onion and many other crops, sodium dichloroacetate and sodium 2,3–dichloroisobutarate (DCIB) have proved quite successful in inducing male sterility. The proper time and dose of application should, however, be determined for individual crops, failing which harmful side effects such as retarded plant growth, malformation of inflorescence, or defective grain set may appear. In addition to this, some unpredictable environmental interactions with the chemical have been reported. Nevertheless, the benefits of producing instant male-sterile lines and eliminating reliance on scarcely available naturally occurring genetic or cytoplasmic male sterility are quite attractive. Further research is required in this area. This technique can be exploited for commercial production of hybrid varieties, particularly when hand or bulk pollination is not economically feasible.

2.3.3 How to determine the mode of pollination

Understanding of the mode of pollination of any crop variety to be selected for genetic improvement is extremely important, because it directly determines the approach a breeder should follow for selfing or making crosses. In a highly self-pollinated crop, like wheat, rice, or barley, it is rather easy to determine the frequencies of cross or self-pollination that may occur under field conditions, even though it may vary between individual varieties and under different environmental conditions. An experiment can be designed using a simple, morphologically identifiable genetic marker, for instance glume colour in wheat. One variety carrying homozygous dominant alleles for red glume colour and another variety homozygous for recessive (white glume colour) are planted side by side in a small plot.

Figure 2.16 shows wheat plants with red glumes (*WW*) and white glumes (*ww*) planted in adjacent rows. Normal pollination occurs, in which pollen is freely exchanged among the flowers of each line. Then seed is harvested from the white glume plants and grown in the subsequent season. The relative number of red glume plants (which will be heterozygous *Ww*) in this population will directly indicate the amount of out-crossing that may have occurred.

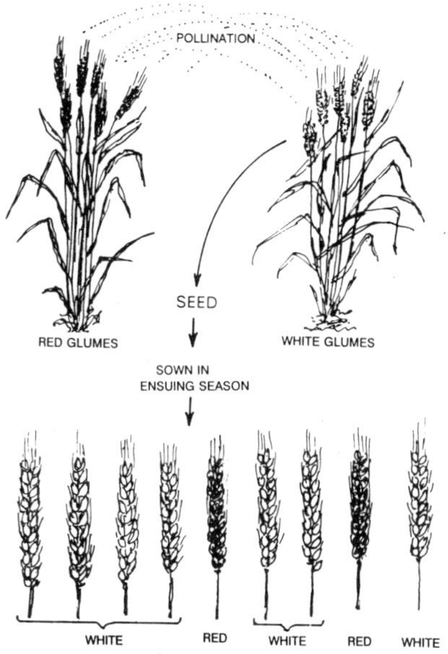

Figure 2.16 Homozygous dominant and homozygous recessive varieties planted side by side to determine natural cross-pollination.

In highly cross-pollinated crops, estimation of the amount of selfing, however, is not that simple. In these types, pollination is normally affected by wind or insects. Some kind of spatial isolation is therefore necessary to prevent plants from receiving foreign pollen. The most practical method is to isolate the flowers with cloth or paper bags. This, however, can cause several other problems; it changes the natural microenvironment around the flowers including light, humidity, and wind; thus the data may not be a true reflection of the natural process. Large cloth cages, into which insect pollinators can also be introduced, are probably better. Varieties differing in simple genetic markers as described before can be grown in these cages along with the insects for pollination. Seed harvested from the recessive parent is subsequently grown and the number of dominant phenotypes in this population will directly indicate the frequency of cross-pollination.

2.5 Reproduction

2.5.1 Sexual reproduction

The sexual process of reproduction involves the fusion of two highly specialized cells called sex cells or gametes. The gametes are always haploid and are carried in the gametophytic generation of plants. Fusion of the male and female gametes gives rise to a diploid zygote, which eventually develops into the embryo contained in the seed. Most lower plants are isogamous; that is, the male and female gametes are morphologically (but not genetically) identical; whereas almost all higher plants are **heterogamous**, that is having morphologically dissimilar gametes. In heterogamous plants, the **megaspore** (egg cell) is much larger and contains more cytoplasm than the **microspore** (male gamete) or sperm, which is almost devoid of cytoplasm. The megaspore resides in the embryo sac, the **female gametophyte**, which in turn is enclosed in the ovule. The microspore is located in the pollen grain (the **male gametophyte**) and is actively mobile, so that during pollination it can travel from anther to stigma in open space and then, during fertilization, quite some distance through the stylar tissue along with the growing pollen tube to reach the egg (Fig. 2.17).

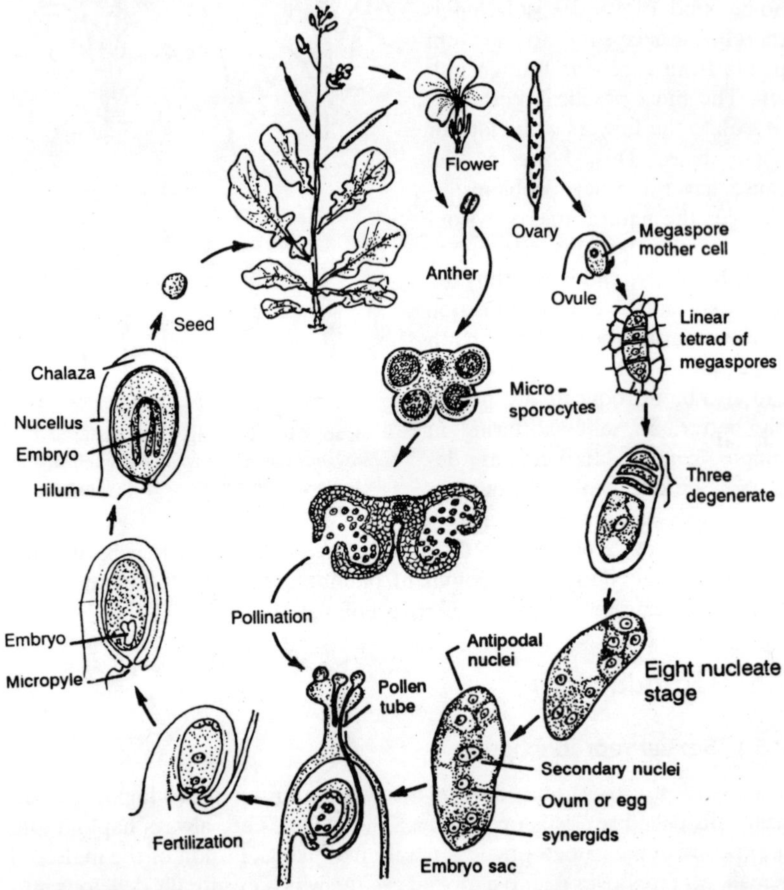

Figure 2.17 Sexual reproduction cycle showing typical stages of the development of male and female sporophytes.

The sexual process of reproduction affords substantial opportunity for variability and blending of genetic characters, a very useful property for plant breeders.

Life cycles. Except for a few instances, all sexually reproducing organisms have a life cycle of two main alternating stages or generations, called the **diplontic** and the **haplontic**. As these names imply, the diplontic stage has twice the number of chromosomes [(2n) per cell] present as at the haplontic stage. In plant breeding terminology this number is usually referred to as diploid, 2n, or the somatic number. The haplontic stage has a haploid number of chromosomes, and is called the n or gametic number. The sexual process itself brings about the combination of chromosome sets through the fusion of two haplontic cells (gametes) to form the first initial cell (zygote) of a new diplont generation. In higher plants the diplont is a large, multi-celled organism (comprising root, shoot, leaves, and flowers), but the haplont is either a single-celled or much reduced structure consisting of only a few cells, as for example the two-celled pollen grain, or (depending upon the species) the 8 to about 40-celled embryo sac (Fig. 2.18). Moreover, these small structures are dependent or parasitic on the sporophyte. In lower plants like the ferns, for example, these two generations alternate during the plant's life cycle, each in turn living independently.

Male and female gametes. In angiosperms, the pollen grains which carry the microspores are produced inside the chambers or **sporangia** of the anther after one meiotic division of the microspore mother cells (**meiocytes**). These microspores differentiate into the male gametophytes (pollen grains), which usually contain one tube nucleus and two sperm nuclei (Figs. 2.19, 2.20). The pollen wall also undergoes further developments. There is a thin inner wall (**intine**) and an outer thick wall (**exine**), on which usually characteristic ridges, spines, grooves, or other forms of sculpturing may appear. On the basis of these features, pollen can often be used to identify the species. During germination the generative nucleus again divides mitotically to give rise to two sperm nuclei which actually bring about fertilization, one of them fusing with the egg and the other with the polar nuclei (Fig. 2.19).

Figure 2.20 shows cross-sections of barley anthers at successive stages of development: (1) early stage: e = epidermis; m = middle layer; pl = parietal layer and central core occupied by al = archisporial cells. (2) primordial stage showing four layers of sporangial wall: e = epidermis; en = endothecium; m = middle layer; and tap = tapetum. The archisporium has differentiated into sp = primary sporogenous tissue. (3) pre-meiotic stage: mc = microsporocytes or meiocytes. (4 and 5) meiotic stages: the meiocytes are at telophase I and tetrad stages. (6) Free microspore stage. (7) Mononucleate state. (8 and 9) mature sporangium with fully developed pollen grains.

The female gametophyte develops in the embryo sac contained in the ovule, a specialized structure of the ovary. Within the ovule (**megasporangium**), a single enlarged megaspore mother cell (megasporocyte or female meiocyte) undergoes meiotic division once, forming a linear tetrad of four

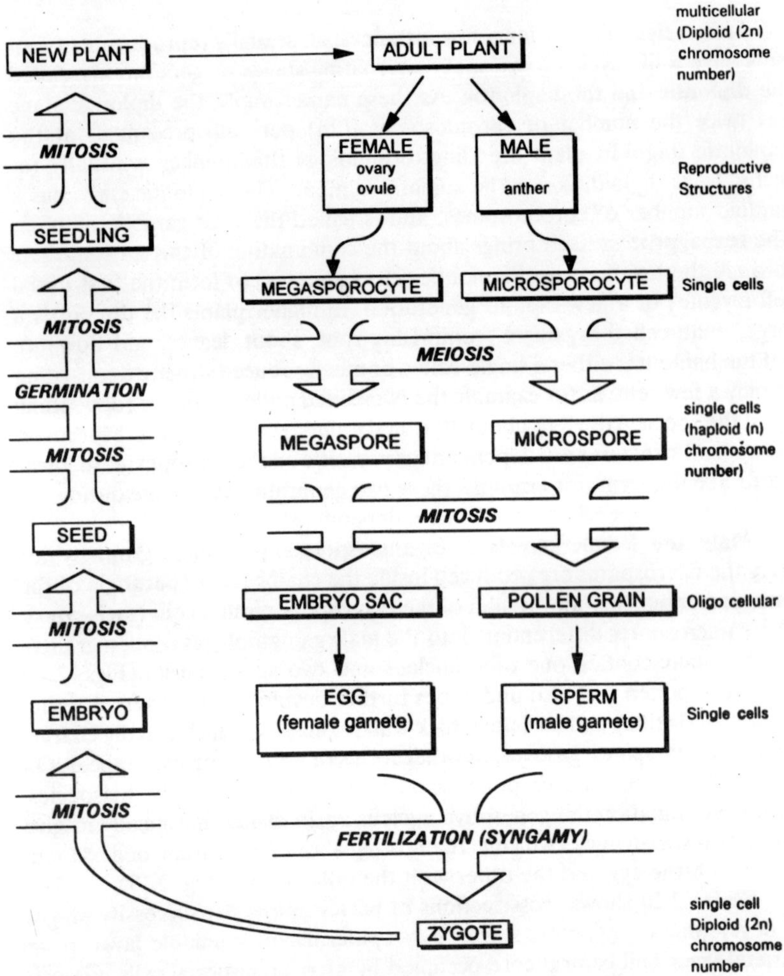

Figure 2.18 Life cycle of a plant

haploid megaspores. Soon after, three of these cells disintegrate, and the one remaining becomes the functional megaspore. Then three successive mitotic divisions occur resulting in eight nuclei. This eight nucleate, haploid structure is the much reduced female gametophyte, called the embryo sac. One of these eight nuclei moves near the micropylar end and differentiates to become the egg cell. Although there are various patterns, depending upon the species, commonly a fully-developed female gametophyte consists of a group of eight haploid nuclei. One nucleus is the egg, which after fertiliza-

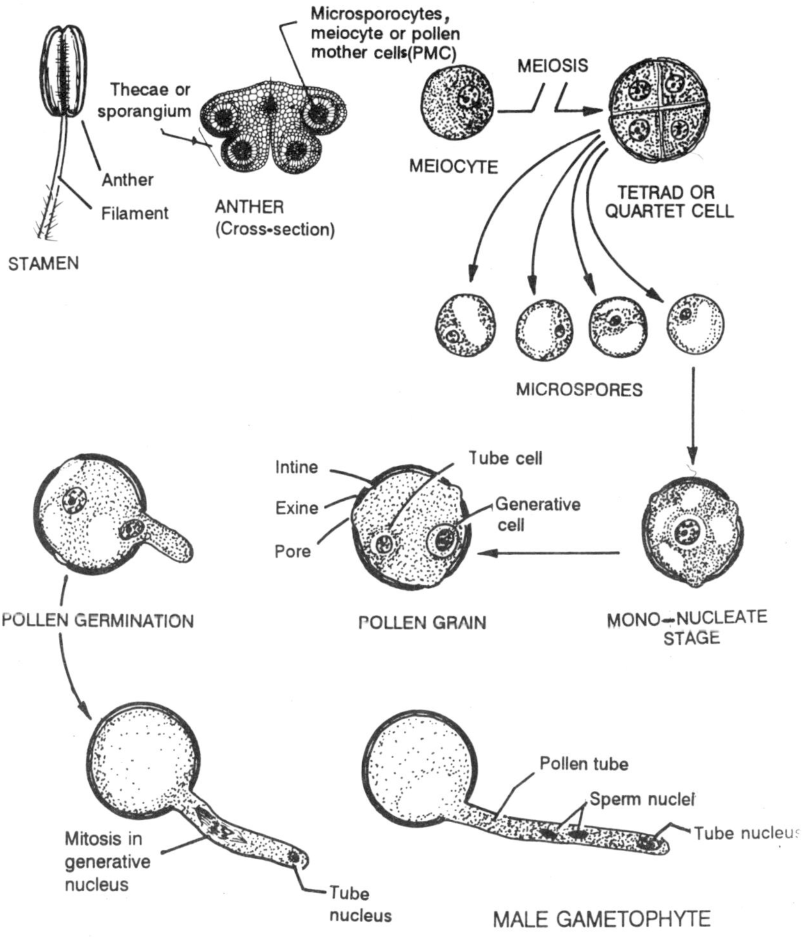

Figure 2.19 Development of male gametophyte (gamete–bearing generation).

tion becomes a zygote and develops into an embryo; two polar nuclei fuse together becoming diploid, and after fusion with a male nucleus become triploid, eventually growing into the endosperm of the seed. Figure 2.21 shows the development of the female gametophyte.

In Figure 2.22, the development of female gametophyte in lily is illustrated with photomicrographs. (1) The lily ovule consisting of one large megaspore mother cell (mc) is in the central position, with well-developed integuments (im) as 'U'-shaped structures covering it. (2) The megaspore

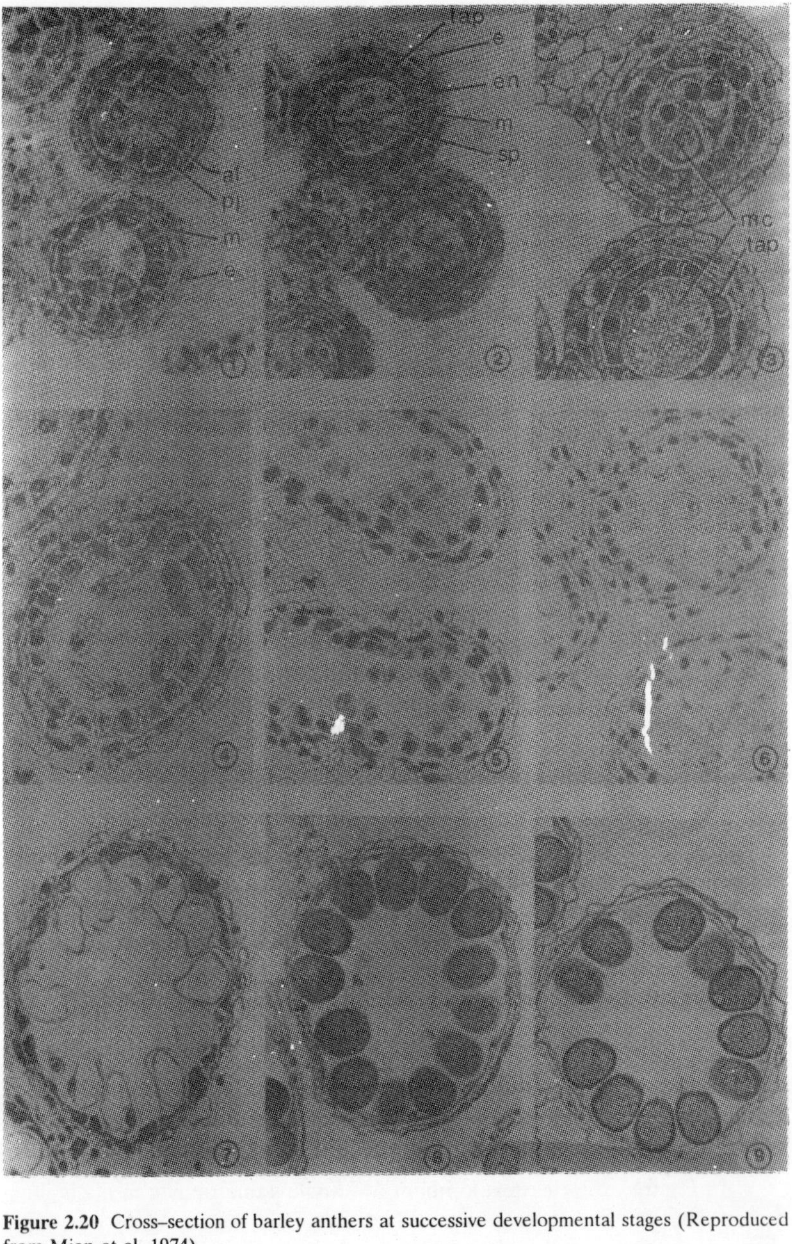

Figure 2.20 Cross–section of barley anthers at successive developmental stages (Reproduced from Mian et al. 1974).

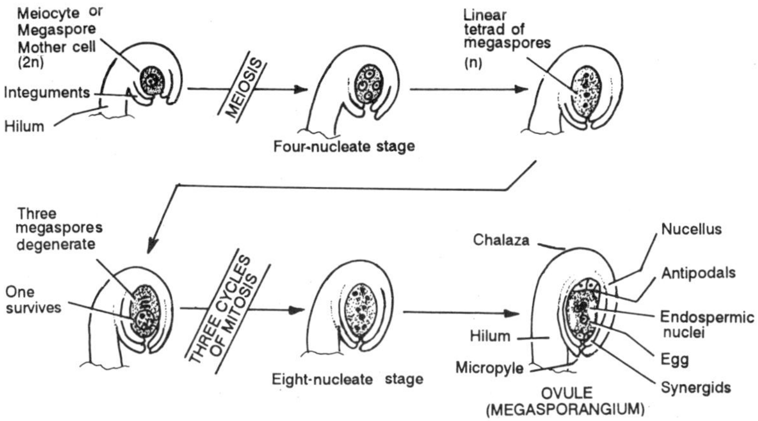

Figure 2.21 Schematic diagram of ovule and embryo sac (female gametophyte) development.

mother cell with conspicuous nucleons (nu) and nucleolii (ni). (3 and 4) Meiotic division: anaphase and telophase I. (5) Late anaphase II. (6) Meiosis is complete and four haploid nuclei are formed. (7) Mitosis in each of the four haploid nucleons. (8) Eight-nucleate embryo sac: three occupying the chalazal pole (called antipodals) and two located near the micropylar end (synergids), while from the remaining three, one is destined to become the ovum and the others the endoplasmic nucleus. These stay in the middle. (9) Mature embryo sac (female gametophyte).

The exact function of synergids and antipodals is not clearly known. The embryo sac assumes ellipsoidal, elongated, or variously bent shapes, depending upon the characteristics of individual species, with its longitudinal axis extending from the chalaza to the micropyle. A constant nutrient supply to the embryo sac is provided through the chalaza, establishing a polar gradient from the antipodals to the micropylar end. Nutrition is also obtained directly through the nucellus and integumentary layers of the ovule.

Development of embryo and seed. The fusion of one sperm nucleus with the egg gives rise to the zygote, and the second sperm nucleus fuses with the polar nuclei to yield a triploid polar or endospermic cell. This is known as **double fertilization**. The zygote grows into the embryo, while the triploid polar cell eventually develops into the endosperm. Following fertilization, the minute embryo sac located inside the ovule grows along with the ovule into a seed. A seed is a ripened ovule, whereas the fruit is a ripened ovary. Often but not always the ovary will not develop unless it contains normal, properly growing seeds. Common exceptions are parthenocarpic seedless grapes, orange, persimmon, banana, seedless watermelon, and cucumber, in which the ovary grows into a succulent fruit without the

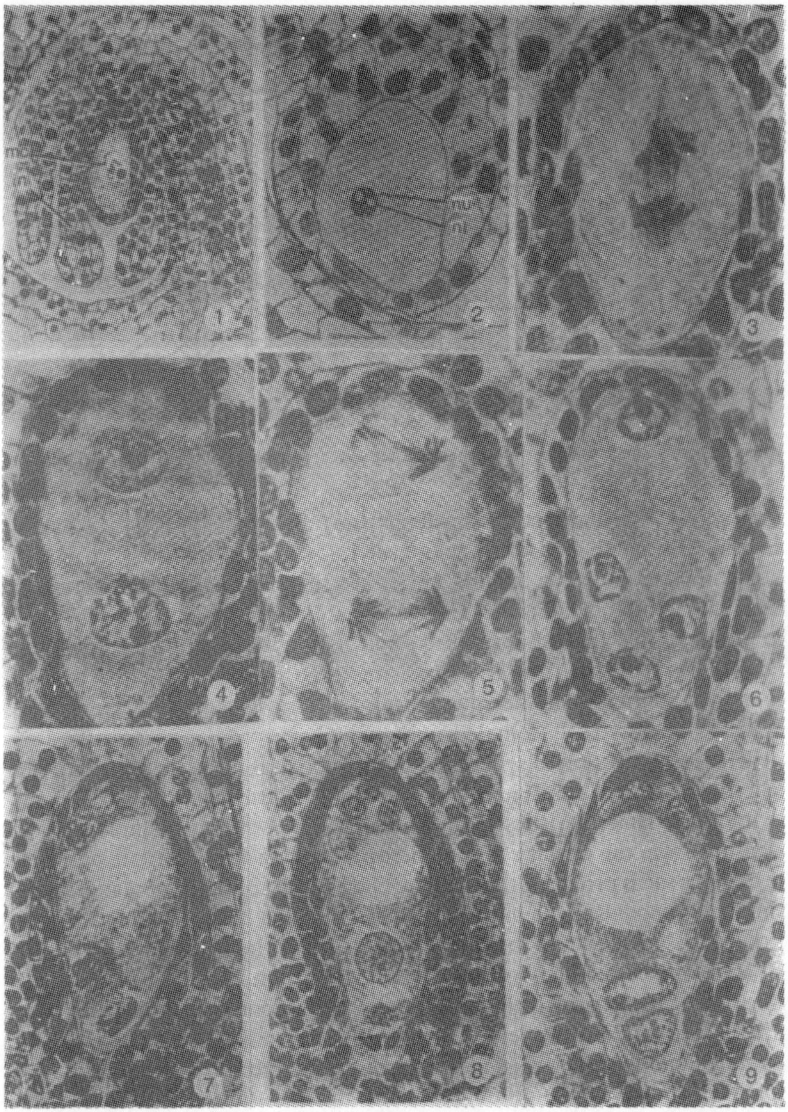

Figure 2.22 Photomicrographs of successive stages in the embryo–sac development in *Lilium*. Source: *Carolina Tips* 11:41–44. Carolina Biological Supply Co., Burlington, N.C. 1973.

formation of seeds.

Most plants exhibit initial similarities in developmental pattern; nevertheless, seed formation is directed according to the genetic specifications of each individual species. In the developing seed, growth of the endosperm is relatively faster than that of the embryo. However, when rapid growth of the embryo begins, it does so at the expense of the endosperm. The size of the endosperm in the mature seed varies with species. In some species it occupies a significant portion of the seed and stores large amounts of food, while in others it is completely absorbed during embryo development and the stored food is kept in the embryo itself, as for example in soybean seed.

Seed development is completed with the formation of a stable, hard,

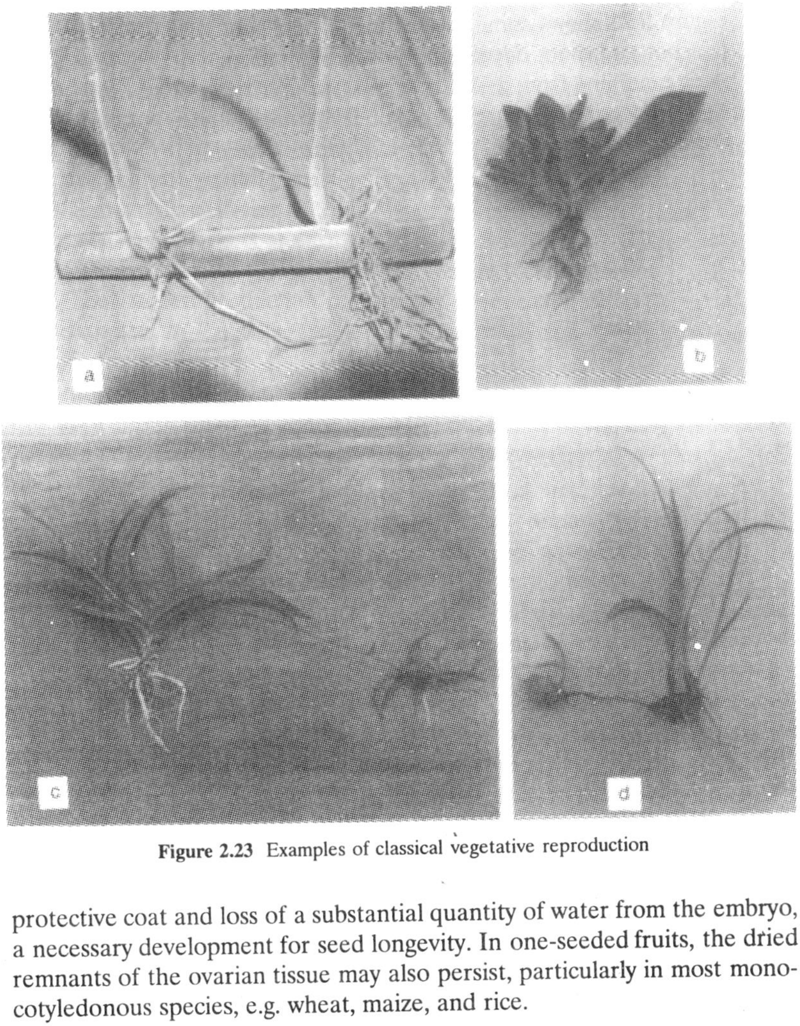

Figure 2.23 Examples of classical vegetative reproduction

protective coat and loss of a substantial quantity of water from the embryo, a necessary development for seed longevity. In one-seeded fruits, the dried remnants of the ovarian tissue may also persist, particularly in most monocotyledonous species, e.g. wheat, maize, and rice.

2.5.2 Vegetative reproduction

Throughout the plant kingdom, including both higher vascular plants and lower forms like algae and fungi, different forms of asexual reproduction are found. Genetically speaking, all types of asexual reproduction are vegetative or clonal in nature. In this sense therefore, the terms *asexual reproduction* and *vegetative reproduction* are synonymous. Botanically, however, these terms are often described separately because of the plant organs or structures involved. According to this classification, reproduction by stem, root, and leaf cuttings or even some floral parts is classed as *vegetative reproduction*. Figure 2.23 shows some interesting examples of vegetative reproduction: (a) root and shoot developing from a stem cutting of sugarcane; (b) a new plant emerging from a leaf of *Echeveria pulvinata*; (c) a bulbil with a new plant growing from a specialized shoot of *Chlorophytum comosum*; (d) a new plant arising from a stem tuber of *Cyperus rotundus*.

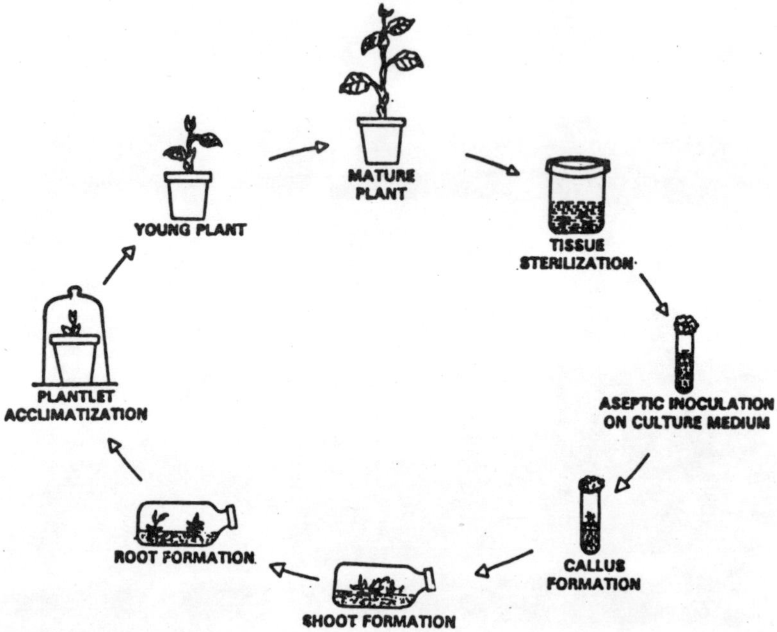

Figure 2.24 Diagrammatic representation of the development of normal tobacco plant by tissue culture from stem pith cells.

The modern technique of tissue culture in which complete plants are regenerated from a protoplast, a single cell, or small group of cells (explant), is also, in fact, vegetative reproduction. Figure 2.24 schematically illustrates the techniques of micropropagation of plants from somatic tissue in an artificial culture medium. This method has now been commercially adapted for mass scale clonal propagation of several crop species. Figure 2.25 shows plantlets generated by this technique: *Brassica napus* regenerated from a callus of hypocotyl (a), and potato leaf tissue grown into a new plant (b).

In addition, many plant species develop peculiar reproductive structures

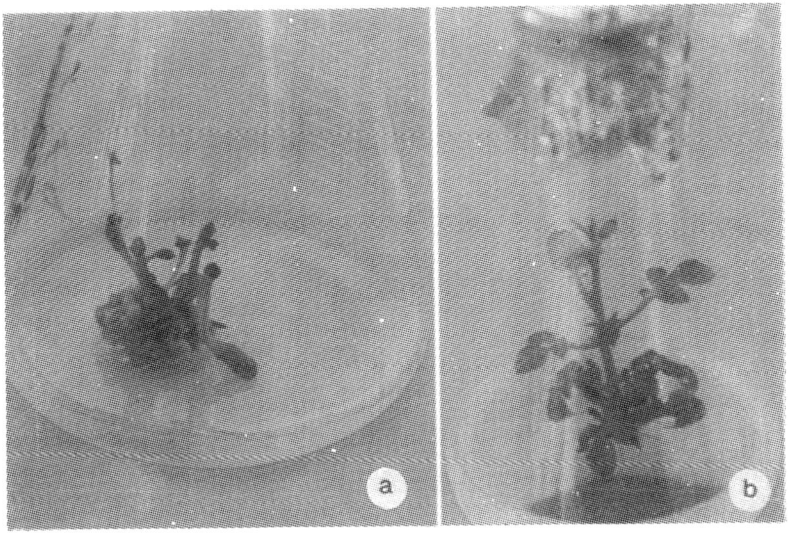

Figure 2.25 Plantlets produced by *in vitro* culture. Photographs courtesy: Said Hassan, Principal Scientific Officer, Nuclear Institute for Food and Agriculture, Peshawar, Pakistan.

that are produced in essentially floral organs, bypassing the normal mechanisms of gamete formation and syngamy. This type of reproduction, although less often found in crop plants, exhibits many variations and is commonly known as **apomixis**. It will be discussed later in more detail.

Advantages of vegetative reproduction. A large variety of economic plants (fruit trees, ornamental plants, some crops) are almost always propagated vegetatively, even though in many cases normal seeds are produced. The main reasons for adopting this procedure are:

1. To maintain genetic homogeneity or keep the crop true to type. The vegetatively propagated plants will be exact genetic copies of the parent even though it is extremely heterozygous.
2. To reduce long generation time. If grown from seed, plants (particularly tree species) take a long time to come to fruiting, but vegetatively grown plants have a relatively shorter juvenile period. Also, seed dormancy problems are avoided.

3. Economic factors. In some crop species, vegetative propagation is routinely used instead of seeds because it is economical, easier, and faster. Even in annuals such as potato and sugarcane, sexual reproduction is used only rarely for new variety development or other experimental purposes, whereas the main crop is always grown vegetatively Sugarcane, for example, flowers only in some areas (Fig. 2.26).

Figure 2.26 Sugarcane flowers profusely under the climatic conditions of Dargai (Malakand) and Chara Pani (Murree).

4. To overcome sterility. Plant species which are genetically sterile or produce no fertile seeds can only be propagated by vegetative means.

There are several methods of vegetative propagation, and the choice depends on the biology of the plant and the motives of the grower. Some of the methods of vegetative propagation are given in Table 2.2.

Table 2.2 Methods of vegetative reproduction for various crops

Plant part	Crop
Shoot cutting	Sugarcane, *bajra* × Napier grass hybrid, large number of forest, horticultural, and ornamental species.
Runner, stolon	Strawberry, kudzu vine, mint.

Plant part	Crop
Rhizome	Lily, iris, banana, numerous species of grasses, ginger, turmeric, colocasia.
Stem tuber	Potato, Jerusalem artichoke.
Bulb	Onion, garlic, tulip, lily.
Sucker	Date palm, banana.
Corm	Crocus, gladiolus, colocasia.
Tuberous root	Sweet potato, dahlia, peony, begonia, asparagus.
In situ grafting, budding, layering (induction of adventitious roots and shoots)	Many fruit trees and ornamental species.
Micropropagation: tissue culture techniques in which small pieces of tissue or even single cells are used.	Successes have been achieved in a large number of plant species.
Embryo culture (growing of excised embryos on artificial nutrient media)	Several crop species can now be propagated.

2.5.3 Asexual reproduction (apomixis)

The term **apomixis** (Gk. 'without mixing') describes several types of non-sexual reproduction in plants. Essentially it is the production of seed or seedlike propagules that emerge directly from the megaspore (unfertilized egg), other cells of the embryo sac, or from the diploid tissue of the ovule. The apomictic process appears to be independent of either fertilization or embryogenesis. More commonly, in addition to the normal sexual production of the embryo (**amphimixis**), one or more asexually created embryos occur in the seed. These have exactly the same chromosome constitution as the mother plants, and the offspring from such embryos of any one plant are true clones. These later embryos can arise from various cells of the embryo sac (unfertilized egg, antipodals, synergids), the inner integuments of the ovule, or the nucellar tissue. There are plants capable of both sexual and apomictic reproduction simultaneously within the same ovule. In such plants, apomixis exists as a deviation along with the normal sexual embryogenesis; these plants are called **facultative apomicts**. They produce two kinds of seeds and eventually offspring: those originating sexually will exhibit Mendelian segregation while the ones of apomictic origin will be identical to the mother plant. In certain other species, apomixis is found as a regular phenomenon, replacing normal sexual reproduction. These are known as **obligate apomicts**. Since the reproductive process inside the ovule cannot be controlled, the breeding of apomictic species poses special problems primarily because their progeny always show unpredictable results. To ascertain the breeding potential of an apomictic species, it is important to understand its exact reproductive behaviour.

2.5.4 Forms of apomixis

In its general features, apomixis is the result of basic disruptive changes in the course of the normal embryological or cytological process. As a consequence, meiosis in megasporogenesis and subsequent differentiation of the egg cell or fertilization may be obliterated. Due to these conditions, the normal formation of actual seed stops and is replaced by the rare non-sexual process that produces pseudo-seeds. Another hypothesis is that some kind of physiological stimulus is given to the surrounding non-zygotic tissue (including synergids, antipodals, nucellus, etc.) which then, deviating from its normal function, develops directly into an embryo. An extensive range of variations of apomixis has been found in various plant species. Since it is not possible to discuss all of them in full detail, only some of the more important ones are described here. Some have also been illustrated in Figure 2.27.

Parthenogenesis. One form of apomixis quite common in plants is **parthenogenesis**, which means development of embryo, seed, or fruit without fertilization of the egg. In this case the embryo is usually haploid. Sometimes, however, such embryos may be diploid or even polyploid because either meiosis has failed to occur or polyploidy is induced during the course of development. Common crop species may show spontaneous, though infrequent, occurrence of parthenogenesis. It can also be artificially induced by hormone treatment or breeding methods, though with poor success. There is evidence that polyploidy and hybridization somehow promote parthenogenesis although they are not the direct cause of it. Probably the heterotic effect or the combination of diverse genomes in hybrids favours the parthenogenetic phenomenon (Burton and Forbes 1960).

Agamospermy. Various disruptions in the development of the egg are collectively referred to as **agamospermy** or **aneuspory**. The seedlike structures or propagules are produced by adventitious embryony, diplospory, or apospory. Pollination may occur but no fertilization takes place, hence the pollen does not make any contribution to embryo formation. Several variants are known, but only two of the most important are described here.

I. In the megaspore mother cell the first meiotic division (the reductional division) takes place normally, but for unknown reasons meiosis II fails to occur. Instead, after a prolonged resting stage, the cells resume formation of the embryo sac, and in this process their chromosome number is doubled. The spores thus formed have a diploid chromosome number and represent only the maternal genotype, possibly also homozygous. These

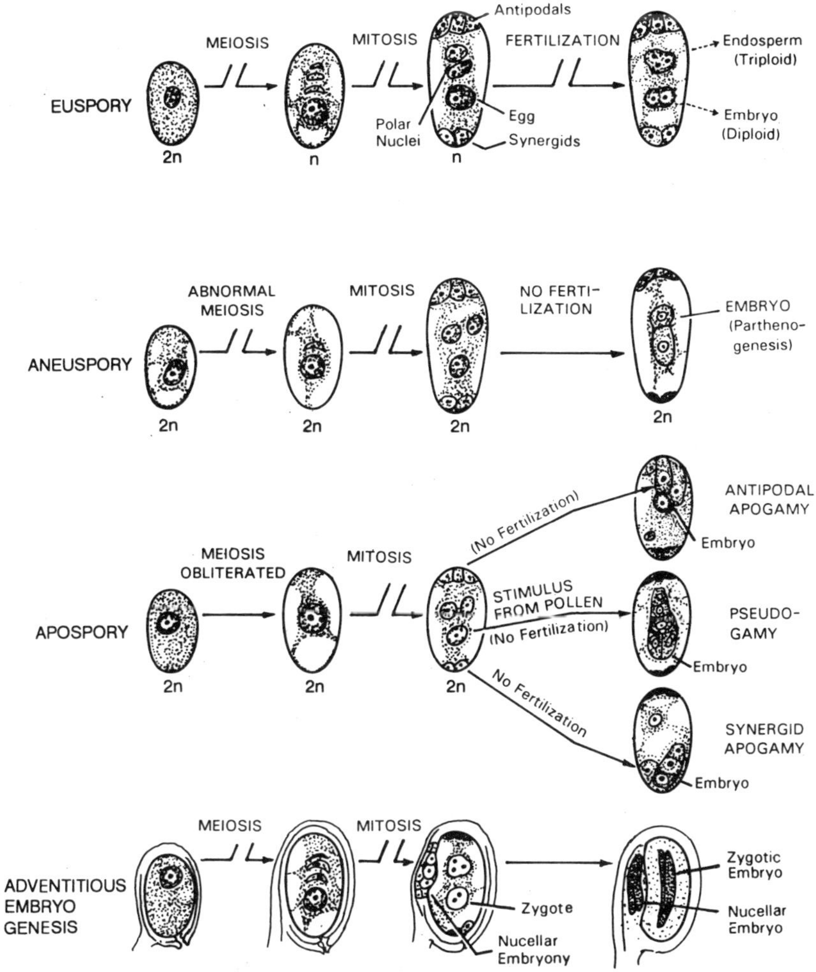

Figure 2.27 Some different forms of apomixis

spores remain unfertilized and develop into the embryo. This is found quite commonly in datura.

II. A rather more extreme case of meiotic deviation is found in certain grasses (*Taraxacum* and *Panicum* spp.). Meiosis is initiated in the megasporocyte, but no synapsis takes place at zygotene, and the chromosomes remain univalent at diakinesis and do not exhibit proper metaphase orientation. Eventually all the chromosomes are enclosed within the same nucleus; that is, instead of two haploid nuclei being formed a single restitutional (unreduced) nucleus results. The second meiotic division may then proceed normally to produce diploid spores.

In other instances (e.g. *Rudbeckia* spp.), meiosis is entirely replaced by mitosis, thus leading to the formation of an unreduced cell in the embryo sac. This cell (egg) ultimately develops into an embryo. In all of these cases, therefore, the embryos arise without meiosis and syngamy, hence they will represent only the maternal genetic constitution.

Apospory. In **apospory**, spores are not produced at all. Meiosis fails to occur and the cells of the embryo sac may arise directly from the megasporocyte (megaspore mother cell)—the cells which will divide by meiosis to produce spores by mitotic divisions (Fig. 2.27). One or more of these cells then differentiate to become embryo(s). Alternately, the embryo develops from somatic cells including parts of the ovule (such as the nucellus or ovular integuments), substituting for the megaspore mother cell. Nucellar tissue, for instance, begins to grow and differentiates into a false embryo sac and eventually an embryo. Such embryos, by virtue of their origin from somatic tissue, consist of unreduced diploid maternal cells. The megaspore mother cell, however, remains independent; that is, it may or may not undergo meiosis and further development.

Pseudogamy. A form of apomixis known as **pseudogamy** occurs due to a stimulus from pollen tube growth. The male gametes enter the embryo sac and one of them fuses with the polar nuclei, but the other one for physiological reasons fails to fertilize the egg nucleus. However, even without fusion with a pollen nucleus, the egg grows into an embryo. The endosperm that develops after double fertilization from the polar nuclei gives a false growth stimulus to the egg, which essentially was unfertilized. In certain cases, synergids or antipodals may form embryos in addition to the one formed from the egg. This is known as **apogamy**. Based on this fact, it is assumed that all the cells present in the embryo sac are potential gametes.

Implications of apomixis for plant breeding. A plant breeder has to be thoroughly aware of the reproductive mechanism of a given species, for it can directly influence the results of a cross. Apomictic plants do not usually offer much problem for hybridization work. Ordinary techniques are applicable to them, and even sexual × apomict crosses can be made. Nevertheless, one should clearly understand that except for rare fertilization an obligate apomictic plant cannot be used as the female parent; it can only be utilized as the pollen parent.

Crosses made in apomictic species will predominantly produce progeny genetically resembling the female parent, because no genetic contribution is made by the male parent. This phenomenon can be exploited to advantage in breeding programmes. For example, a superior variety which reproduces by apomixis will breed true for maternal characters, eliminating the risk of genetic segregation due to outcrossing which would be present if seeds were produced by sexual means. In this manner good phenotypes can be kept stable once they have been isolated. Similarly, highly heterozygous varieties (particularly fruit species) can be cloned by apomixis without disturbing the phenotypic qualities of the variety.

On the other hand, apomixis also has some disadvantages. It inhibits genetic variability (which normally originates by segregation or recombination), causing a problem for the breeder when he is interested in segregating genotypes in plant progeny. Moreover, in cross-pollinated crops frequent appearance of maternal-like phenotypes due to apomixis causes confusion and erroneous segregation ratios. Under such circumstances, genetic markers have to be used to identify the frequency of apomixis being exhibited in the progeny.

STUDY QUESTIONS

1. Botanically, a flower is composed of whorls of highly modified leaves which are the calyx, corolla, androecium, and gynoecium. What specific functions does each of these whorls perform?
2. Flowers are the organs of reproduction in plants. Why do we observe such great variation of pattern, composition, and physiological behaviour in flowers?
3. Given the fact that plants can reproduce by vegetative means, why do you think there is a need for sexual reproduction? Support your answer with examples.
4. What are the principle cytogenetic features of sexual reproduction?
5. What are the advantages and disadvantages of sexual and asexual reproduction methods with respect to plant breeding?
6. Illustrate with diagrams the life cycles of cotton, potato, and *Neurospora crassa*. Tell which of the phases—haplontic or diplontic—is more conspicuous and/or long-lived, and state the somatic chromosome number in each phase.
7. Define the relationship of meiosis and syngamy to the haplontic and diplontic phases.
8. What is the significance of meiosis and mitosis in crop improvement?
9. Explain the term *clone*. Under what conditions does a plant breeder need to adopt clonal propagation?

10. Describe the ontogeny, development, and function of endosperm. List the crop species that have seeds with well-developed endosperm.
11. In maize (*Zea mays*), a monoecious species, the 3n endosperm may be flinty (F) or floury (Fl). Reciprocal crosses between true-breeding flinty and true-breeding floury strains produce the following results:

Parent			**Endosperm type**
♀		♂	
F	×	Fl	F
Fl	×	F	Fl

(a) Explain why reciprocal crosses give different results. (b) Draw a diagram showing all nuclei of ♂ and ♀ gametophytes, indicating the genotype before and after fertilization.

12. Explain the difference between cross-breeding, in-breeding, and sib-breeding, and show what type(s) can occur in (a) unisexual (dioecious) species; (b) bisexual (monoecious) species.
13. Would you expect higher phenotypic differences among progeny of sexual or asexual reproduction? Explain the reasons for your answer.
14. Do you think that genetic variability is an advantage or disadvantage for the survival of a species under natural conditions? Explain, with reasons and examples.
15. What different mechanisms cause male sterility? Is male sterility in crop plants an advantage or disadvantage from the breeding point of view?
16. Explain with diagrams how genetic and cytoplasmic male sterility can be sustained in the progeny of a crop species.
17. Assume that a strain of male-sterile maize is discovered by a breeder at your university. Suggest a method for ascertaining whether the trait is controlled by cytoplasmic or by nuclear genes.
18. Describe how deletions, translocations, and inversions affect meiosis.
19. Some types of chromosome aberrations, deletions, translocations, and inversions cause semi-sterility in plants. Explain how each of these conditions induces mortality in gametes.
20. In onion, F_1 hybrids give very high yields, but because of the minute size of the flowers, crossing is extremely difficult. To overcome this problem, male-sterile lines can be used for hybrid seed production.

Suppose that a male-sterile line is discovered. Crosses made between this male-sterile line and some fertile ones give the following results:

	Parents			Progeny
	♀		♂	
1.	Male-sterile	×	Male-fertile (Variety 1)	All male-fertile
2.	Male-sterile	×	Male-fertile (Variety 2)	All male-sterile
3.	Male-sterile	×	Male-fertile (Variety 3)	1 male-fertile, 1 male-sterile

Analyze these results and suggest whether the male-sterility character is cytoplasmic or genetic.

21. In maize:

Bm = green versus bm = brown midrib.
S = soft versus s = brittle endosperm
G = green versus g = virescent seedlings.

A cross between bmbmssgg as ♀ × BmBmSSCG as ♂ in F_1 produced plants showing a dominant phenotype for all three characters, but one plant in the progeny expressed all three recessive traits. This plant also had shrivelled and defective pollen, and at pachytene also showed a short buckled region in the long arm of chromosome 5. Explain these results giving reasons for your answer. How could a plant with all three recessive traits appear in F_1?

22. Sexual reproduction requires the formation of haploid gametes and their fusion to produce a diploid zygote. This in turn develops into an embryo and eventually a new plant. Answer the following questions regarding this process. (a) Differentiate between microsporogenesis and megasporogenesis. (b) What specialized cell undergoes meiotic division in each of the above? (c) Which organs of a flower grow into seed and fruit? (d) Name the two distinctly different structures that develop after double fertilization; what is the genetic makeup and function of each? (e) Which cell of a mature gametophyte usually develops parthenogenetically?

23. In *Nicotiana*, if S^1S^2 is crossed with S^4S^5, what will be the proportion of wholly or partially incompatible combinations among the crosses of F_1 individuals?

24. If a plant heterozygous for one allele pair *Aa* is continuously self-pollinated, what proportion of the progeny will be *Aa* after seven generations?

25. Assume that a plant flowers and sets seed during the months of March–April and is naturally adapted for cross-pollination and heterozygous for one allele pair *Aa*. It is allowed to freely cross-pollinate in an isolated place. After two seasons of breeding, what proportion of the progeny will be homozygous recessive (*aa*), homozygous dominant (*AA*), and heterozygous (*Aa*)?

REFERENCES

Ahmad, S. S. A textbook of intermediate botany. 26th ed. Lahore: Sheikhsons Publishers; 1984.

Burton, G. W.; Forbes, I. Jr. The genetics and manipulation of obligate apomixis in common bahia grass (*Paspalum notatum*). In: Proc. 8th Intl. Grassland Congr. held in Reading, UK, 11–21 July, 1960. Oxford (UK): Alen Press; 1960:p. 66–71.

Elliott, F. C. Plant breeding and cytogenetics. New York: McGraw–Hill; 1958.

Frankel, R.; Galun, E. Pollination mechanisms, reproduction and plant breeding. Monographs on Theoretical and Applied Genetics, Vol. 2. Berlin: Springer–Verlag; 1977.

Hans, T. E.; Ramage, R. T.; Tsuchiya, T. Description of genetic stocks in the barley genetic stock center at Fort Collins, Colorado. In: Barley Genetics Newsletter 1:175–89. Fort Collins, CO: Department of Agronomy, Colorado State University; 1971.

Hockett, E. A.; Eslik, R. F.; Ried, D. A.; Wiebe, G. A. Genetic male sterility in barley. II. Available spring and winter stock. Crop Sci. 8(6):754–55; 1968.

Jain, S. K. Male sterility in flowering plants. Bibliogr. Genet. 18:103–166; 1959.

Khan, M. D.; Rahman, H. Genome relationship and chromosome behaviour in the allotriploid hybrid of *Pennisetum typhoides* and *P. purpurum*. W. Pak. J. Agric. Res. 1(3):61–65; 1963.

Khush, G. S. Cytogenetics of aneuploids. New York: Academic Press; 1973.

Mian, H. R.; Kuspira, J.; Walker, G. W. R.; Muntjewerff, N. Histological and cytochemical studies on five genetic male-sterile lines of barley (*Hordeum vulgare*). Can. J. Genet. Cytol. 16(2):355–79; 1974.

Mian, H. R.; Kuspira, J.; Walker, G. W. R.; Muntjewerff, N. Macromolecular changes and submicroscopic structure in the differentiation of msg 16 male-sterile barley anthers. In: Barley Genetics 4. Proceedings of the 4th International Barley Genetics Symposium, Edinburgh, 22–29 July 1981. Trowbridge, Wilts., UK: Redwood Burn, Ltd. (Edinburgh University Press); 1981:p. 804–13.

Rahman, H. Anthesis, pollination and qualitative studies on Henbane (*Hyoscyamus niger*). Lahore: University of the Panjab; 1952. M.Sc. thesis.

Rick, C. M. The tomato. Scientific American 239(2):66–78; 1978.

Roath, W. S.; Hockett, E. A.; Pollen development in genetic male-sterile barley. In: Nilan, R. A. ed. Barley Genetics II. Pullman, Washington: Washington State University Press; 1971: p. 308–315.

Swanson, C. P.; Merz, T.; Young, W. J. Cytogenetics: The chromosome in division, inheritance, and evolution. Englewood Cliffs, NJ: Prentice–Hall; 1981.

Outline of Chapter 3

3. BREEDING METHODS FOR SELF-POLLINATED CROPS

Khurshid Alam[1]

LEARNING OBJECTIVES

After studying this chapter, a student should be able to:

- Explain the mechanics of the breeding or selection methods generally employed in the breeding of self-pollinated crops like wheat.
- State the defining characteristics of each method, and tell how each of the methods differs from the others.
- Given a specific plant breeding problem, decide correctly which plant breeding method is most appropriate for that situation.
- Justify a specific plant-breeding decision based on fundamental principles of plant breeding science.

CHAPTER OVERVIEW

Naturally self-pollinated crops possess flower structures containing both male and female parts which show perfect compatibility, a situation which leads to self-fertilization. This mode of pollination and fertilization ultimately results in homozygosity. Thus plant species where self-fertilization takes place as a rule contain all homozygous plants, provided that these populations are allowed to propagate through generations uninterrupted by outside forces such as man-made and natural crosses, or mutations.

In these populations, however, a certain amount of crossing (hybridization) does occur in nature. This introduces a degree of heterozygosity among them, which is the prime source for change in their gene frequencies, and a basis for selection. This means that self-fertilized crop species, with very few exceptions, are not 100 percent homozygous; a small fraction of the population is always heterozygous.

---

[1]Dr. Khurshid Alam is Professor and Chairman of the Department of Plant Breeding and Genetics, University of Agriculture, Faisalabad.

Breeding methods designed and usually adopted for the improvement of self-fertilized crop species like wheat, barley, or oat envisage the development of plant populations consisting of homozygous plants. Such a crop variety can remain under cultivation in a homozygous condition for a long time before it runs out, because of mechanical mixing of seed of different varieties over the years or because of natural crossing and mutation. Breeders are, therefore, advised to constantly remain on the lookout for variant plants and to take measures to rogue them in order to prolong the life of their cultivar.

The following methods employed in the breeding of self-pollinated crops are discussed.

1. Introduction
2. Pure-line selection
3. Mass selection
4. Hybridization
   a. Pedigree method
   b. Bulk method
   c. Modified pedigree and bulk method
   d. Single-seed descent method
   e. Backcross method

3.1 Introduction

Introduction is a means of benefiting from breeding products emerging at other locations. Finished varieties, strains, or breeding material at advanced stages developed at international centres of crop improvement may be acquired through the good offices of national embassies and various international agencies. They are duly tested for their adaptive potential under local conditions. Of these, some may be found quite suitable for cultivation in our conditions, as they show a good degree of acclimatization. After field trials conducted over varying locations and several years, such materials may be recommended for various localities for commercial cultivation. All countries of the world in the earlier stages of their agricultural development have taken advantage of one another's efforts in this field, of course from areas with more or less similar geographic as well as topographical range. Most of our own crop varieties of wheat, cotton, sugarcane, maize, clover, oilseeds, and several other crops came to this country via this route. In the case of wheat, important varieties like 'Mexipak', 'Sonora', 'Pitic', 'Yecora', and 'Wl–711' have been adopted as our own varieties through introduction.

3.2 Pure-line selection

This is the simplest of all methods, and has been in use ever since man learned to grow crops. A **pure line** may be defined as a population all derived from a *single self-fertilized plant*. The pure-line method is practised in three stages.

Stage 1: Single plant selections are made from the unimproved material in as large numbers as the resources of the breeder permit.

Stage 2: From the seed of each selected plant, progeny rows are grown. A progeny row may be grown from an ear, e.g. in the case of wheat, using 15–20 seeds. The better performing rows are identified and selected, and those not showing good results are discarded. This decreases the number of single-plant selections, leading to better handling.

Stage 3: The selected lines are yield-tested in replicated trials. Those found to be high-yielding are selected finally and approved for further tests, if necessary, before they are released for commercial cultivation. In the early period of our agricultural history, most of the wheat varieties as well as varieties of other crops were developed by the pure-line method of selection.

The method of pure-line selection is diagrammed in Figure 3.1.

3.3 Mass selection

Like single-plant selection as discussed above, mass selection involves selection from unimproved material, land varieties, or any variable source. However, a large number of good-looking plants are selected and bulked to produce the source material for the next cycle of selection. The primary purpose of mass selection is to purify an otherwise adapted variety or to improve an existing variety in some respect, e.g. plant maturity or disease resistance. Since the selected population is generally well-adapted, it requires a minimum of yield testing before it can be released. The mass-selection method ensures retention of sufficient variability in the material being improved or purified to provide genetic flexibility for future selection when needed.

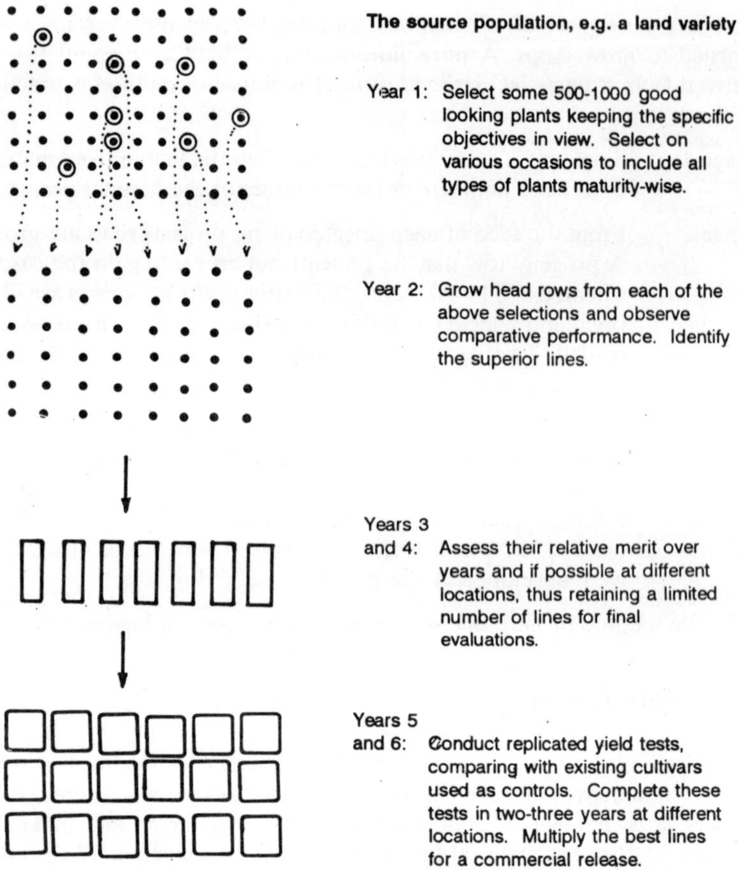

Figure 3.1 Illustration of pure–line selection

3.4 Hybridization

If sufficient genetic variability is not available for effective single-plant or mass selection, the breeder is advised to artificially create variability by making crosses or inducing mutations. The populations resulting from such operations are heterozygous, and may be handled by the pedigree and bulk methods. In this way desirable plant characteristics scattered among different varieties or species can be put together into a single variety.

The choice of parents for crossing thus assumes great importance. It should be dictated by knowledge of their relative merit, which the breeder must manage to acquire. The varieties used in crosses may be locally adapted varieties, exotic varieties, introductions, collections, or species. The crosses that the breeder makes may be classified as safe (obvious) crosses, speculative crosses, or highly speculative crosses. The greater the number of plants used in these crosses, the greater are the chances of obtaining the desired combination of characters and the easier the development of a variety with good agronomic qualities. However, the number of crosses to be attempted by a breeder will depend on the financial and physical resources of his breeding programme.

3.4.1 How to make a cross

Emasculation. In self-pollinated crop plants, if the flower is not prepared for crossing at the proper stage, it will self-pollinate and set the selfed seed. Therefore, the first essential step towards crossing is to emasculate the flower of the plant chosen to act as the female parent. This should be done at the right stage when the anthers have not yet opened and are still green. The emasculated flower should be properly covered with a paper or cloth bag. Glassine or cellophane are good materials for the bags, but polythene bags are to be avoided because they do not permit the flower to have access to air. The bag should remain intact until pollination is done by the breeder. For this purpose, mature pollen is taken from the plant chosen to act as the male parent. The number of flowers emasculated depends upon the breeder's need as well as his resources. These crosses are single crosses, although double, three-way, or multiple crosses may be accomplished in future generations as may be decided in the best interest of the breeding programme.

Pollination. In the case of wheat, pollination is done two to four days after emasculation. Fully mature but as yet unopened anthers at the pre-anthesis stage are collected in a small dish and are applied to the stigma with the help of a forceps or a small artist's brush. For wheat, our most important self-pollinated grain crop, a third method is considered convenient. A male spike at the suitable stage is taken and the top one-third of the florets are cut off so that the anthers are exposed. This spike is then inserted in the inverted position into the bag covering the female flower. For this purpose the top of the bag is cut off and the spike is twisted by rolling the peduncle of the ear between the thumb and the forefinger. This spike may be left in the bag or removed. In yet another method, the male spike is placed in the bag side by side with the emasculated spike. The male spike should be placed slightly higher than the emasculated one to facilitate the pollen falling down onto the stigma as the anthers dehisce.

One precaution is very important in order to get flowers at the right stage for crossing. The parental plants should be grown on different dates at appropriate intervals. This will help synchronize the opening of the flowers of both the parental plants and facilitate successful crossing. A good measure of success in hybridization (crossing) is achieved if the aforementioned operations are carried out carefully. At harvest the crossed (hybrid) seed should be removed from the spike and put in safe custody. Necessary information regarding the number of flowers pollinated and the number of seeds obtained, etc. should be duly recorded on the accompanying tags and field notebooks.

F_1 hybrid generation. The F_1 generation is grown from the hybrid seed obtained from the cross. This may be grown in the field in spaced rows, or in pots in a greenhouse. In the field, for wheat, the rows should be spaced 20–30 cm apart, and the plant-to-plant distance within rows should be 15–20 cm. Likewise, the parents are also grown nearby, preferably surrounding the F_1 rows for easy comparison of the F_1 plants with the parental plants. This is done to establish the hybrid constitution. The F_1 plants which look like the female parent and apparently result from selfing are discarded and eliminated from the nursery. The fully established hybrid plants are marked for record keeping and taking observations. These genuine F_1 plants provide the seed for the following F_2 generation.

F_2 generation. The F_2 generation is the first segregating generation, and from this generation onwards the segregating material may be handled by the pedigree method or the bulk method. We shall first discuss here the important steps involved in the pedigree method.

3.4.2 Pedigree method

The most commonly used method for handling the F_2 and subsequent generations in self-pollinated crops like wheat is the pedigree method. In this method selection starts in F_2. The F_2 generation is planted in spaced rows, growing 2000–10,000 plants depending upon the width of the cross and the resources of the breeder. Genetic variability is at its maximum at this stage, and is reduced by 50% in each successive generation. By the F_5 generation, homozygosity increases to 94%, and by the F_8 generation to 99%.

F_2 plants can be selected from space-planted populations by visual inspection by the breeder in accordance with his objectives. Selections for simply-inherited characters and those with high heritability like disease resistance is easy and quite effective; while for complexly-inherited characters like yield, selection by visual inspection may not be very effective. This makes progeny tests necessary before single-plant selection for such characters can be made. Such selection thus has to wait until later generations.

Plants not conforming to the breeder's objectives should be ruthlessly discarded; otherwise the breeder's workload will be unnecessarily increased, and confusion may result. Some crosses are more vigorous and promising than others and should be concentrated upon. Other crosses will be found to contain poor segregants and should be totally discarded.

About 50–100 F_2 individual plants (1–10% of the F_2 population) should be selected, depending upon the resources of the breeder. However, he must select at least 50 plants. In the case of wheat, the heads of the selected plants should be tied together in bunches of 25–50 heads, and all the necessary information about them recorded on tags. These heads are allowed to dry before they are individually threshed. Ears yielding unattractive grains should be discarded at this stage also.

F_3 generation. Individually selected F_2 plants are grown as F_3 lines/progenies/families in space-planted rows. These rows may contain 20–30 plants along with checks planted at suitable intervals among the rows for ready comparison. Checks may include one or two commercial varieties. Selections are now made on the basis of family performance, and notes should be taken and recorded and kept in the proper notebook. Families/lines are compared visually among themselves and the checks. The better looking families are marked for further selection of individual plants from within each such family.

Selection in the F_3 generation is more drastic than in the F_2, as the family performance is a better indication of their genetic potential and ultimate agronomic merit. The number of plants selected should be less than the number of F_3 lines. This means that many F_3 lines will be rejected on the basis of their poor appearance, and not considered further for the purpose of breeding. The superior families of the plants that have not been selected can be bulk harvested for conducting various tests on their disease resistance and other characteristics.

F_4 generation. Seed of plants selected from F_3 lines or families is grown in space-planted rows as before to produce the F_4 generation. F_4 lines/families will show greater differentiation, and selection will again be made on a single-plant basis from within the most promising families. The number of lines to be advanced to F_5 is drastically reduced by rejecting all lines not meeting the standard. In a similar way, selection should be continued for F_5 to F_7, selecting plants from within the best families, and from their seed raising space-planted families. At the approach of the F_7 generation, near complete homozygosity prevails. Intra-family variation attributable to genetic differences is negligible, and all variation that remains within line/families is environmental. From now onwards, therefore, the lines can be bulk harvested as new strains, which should be tested for various

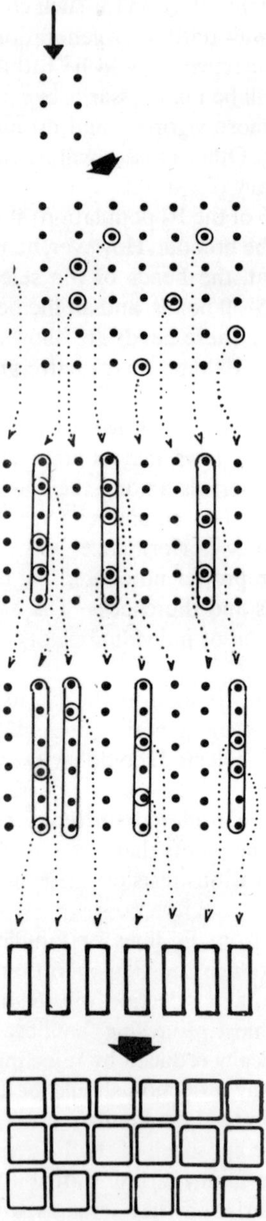

Make the cross between the chosen parents, A and B.

Grow F_1 in space-planted rows and identify the genuine F_1's.

Grow F_2 from genuine F_1 (hybrid) plant seed and select desirable single plants by visual selection.

Grow space-planted F_3 progeny rows. Identify the best lines by visual inspection of individual plants within rows and select the best of them. Reject the poor looking rows (families).

Continue as in the F_3 generation from F_4 to F_7. Surplus seed can be used for quality tests.

In F_8 conduct replicated yield trials.

From F_9 to F_{12} carry out multi-locational replicated yield trials and disease reaction tests and make final selection of the lines to be approved for commercial use.

Figure 3.2 Illustration of the pedigree method of selection characteristics from F_8 to F_{11} and F_{11} to F_{13} as may be decided by the breeder. Such tests are conducted in replicated trials at various places where the varieties are likely to be released for commercial cultivation.

The pedigree method of selection is diagrammed in Figure 3.2.

3.4.3 Bulk method

Unlike the pedigree method described above, in the bulk method the progeny of a cross is handled in bulk instead of on the basis of single-plant selection. No selection is made in the F_2 generation, but a random sample is taken and the bulked seed thus obtained is used to grow the following F_3 generation. In the same manner, the process is repeated to produce the F_4 generation and kept going up to the F_6 generation, when the plants have become nearly homozygous. At this stage, the next generation is space planted and individual plant selection is done. These progenies are homozygous and therefore qualify for bulking to constitute new strains. These strains are handled in the same way as in the pedigree method. Progeny tests are carried out in F_7 lines and repeated over localities and time to collect necessary data on their agronomic performance. Their suitability for commercial use is determined on the basis of these data. Accordingly, new varieties which pass all the field tests with good performance are released. Breeders use the bulk method where a large number of crosses have to be handled or where more complex multi-parental crosses are involved. The handling of such material by the bulk method is also less expensive than the pedigree method because there is no need to keep the complete record of every plant selected from F_2 to F_6–F_7. The bulk method is diagrammed in Figure 3.3.

3.4.4 Modifications of the bulk method

Both the bulk and the pedigree methods have been modified differently by different breeders. We shall talk here about the most commonly used modified versions in crop breeding programmes. A usual modification of the bulk method consists of subjecting the F_2 generation and later generations to artificial disease epiphytotics and other environmental stresses. Under such pressures the undesirable genes are eliminated, and the frequency of desirable genes increases. The probability of selecting better performing plants under such stresses improves greatly.

Single-seed descent (SSD) method. This is another modified form of the bulk method and currently is widely used in breeding self-fertilized crops. The primary purpose of this method is to accelerate the pace and progress of breeding, which is achieved by rapidly advancing the breeding material from the F_2 through the F_6 generations. During this stage, no selection is made. By the approach of the F_6 generation, the plant material

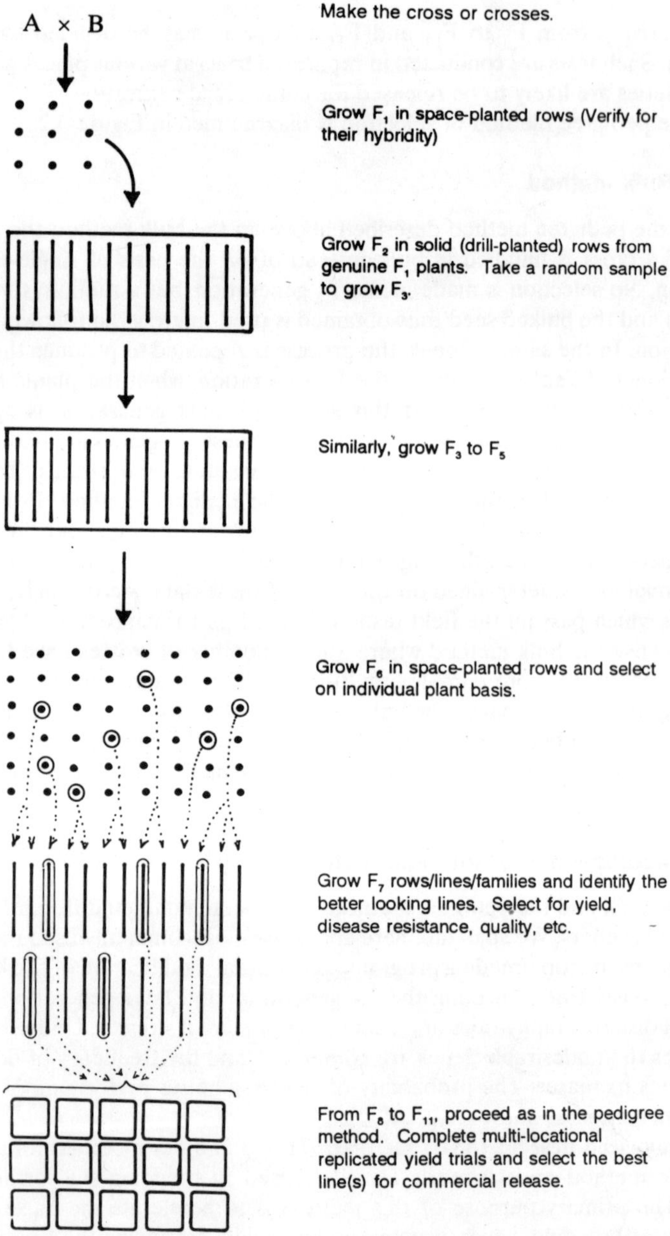

Figure 3.3 The bulk method

has attained near homozygosity. It is at this stage that selection is made on an individual plant basis. Thereafter all operations involving field and laboratory tests are carried out in the same way as in the bulk method.

The method consists of taking one or two seeds from each F_2 plant of the cross under study. The seeds are bulked and sown at a high density (drill planted) on a field with low fertility status. At maturity, one seed per plant is taken. The seed is bulked as before and used to raise the next generation. This process is thus continued until (F_6) selection is made of individual plants from a space-planted population. Head rows/progeny rows are grown in the next generation, to be assessed for their family performance in grain yield and other observable characters. The most promising line/families are identified, from which the best looking plants are again selected. This, however, appears unnecessary, as intra-line variability has by now been completely exhausted.

The best lines are multiplied for location × years yield tests to identify the most desirable line or lines for commercial cultivation.

Advantages of the SSD method.

1. A large number of pure lines developed in a random manner are maintained, permitting better exploitation of the potential of the cross.
2. The full range of variability of the cross is retained for selection in F_6.
3. From F_2 to F_6 the material is afforded ample opportunity for recombination to take place. This breaks undesirable linkages and helps put together desirable genes scattered around in the population.

3.4.5 Modifications of the pedigree method

The most commonly used modified versions of the pedigree method described in the literature are as follows.

1. Early generation testing. In view of the complexity of the genetic control of characters like grain yield which have low heritability, much larger samples of F_2, F_3, and F_4 generations should be grown than is currently done. In addition, much greater attention should be given to the evaluation of F_3 to identify the best lines.

2. F_2 progeny method. F_2 populations of several crosses containing 2000–3000 plants per cross are grown in space-planted rows. For wheat, single-plant selections are made for characters that can be easily identified, such as disease resistance, maturity, grain colour, lodging, etc., selecting 10% of the F_2 population. In F_3, selected progenies are harvested in bulk and reselected on the basis of their family performance. F_4 lines derived from them are grown in replicated yield trials. This procedure is repeated on larger plots and at different locations until in the F_7 generation mixtures of homozygous plants are available. From these mixtures single plant selections are made and retested from F_8 to F_{10} to provide finished varieties.

3. Pedigree trial method. Essentially this is not different from the conventional pedigree method, except that yield tests are concurrently made on selected bulks, beginning from F_4 through F_7. Single-plant selections are made in F_2, and the best families/progenies for F_3 and F_4. After selections have been made, the remaining plants of these families are bulked to provide enough seed for yield trials. This process of single-plant selections and yield trials on the selected bulks is continued until F_7. Each year the seed is derived from selected progeny rows, and not from previous bulks, which are used only for yield testing. Promising lines having been finally identified, replicated yield tests are carried out as usual for three or four years before a finished variety is made available for commercial release.

3.5 Backcross method

This is a relatively inexpensive and straightforward method of varietal improvement. It seeks to bring about genetic improvement by incorporating desirable simply-inherited characters from a close or distant relative into an otherwise desirable commercial crop variety. In this case, the parent donating the desirable gene is called the donor or non-recurrent parent, and the commercial variety receiving it is called the recipient or recurrent parent. This method has been extensively used in the genetic improvement of field crops, particularly in wheat where disease resistance (e.g. rusts) has often been incorporated into commercial varieties by this method.

The operational features of this method are schematized in Fig. 3.4. When the gene being transferred to the recurrent/recipient parent is dominant, the backcross method is relatively faster, because the resistant plant is easily recognizable and there is no need to grow the F_2 generation. After the donor × recipient parent cross, the F_1, which is resistant, is backcrossed to the recurrent parent. This is called the first backcross. The resulting progeny of this backcross will consist of 50% resistant (heterozygous) and 50% susceptible (homozygous) plants. The heterozygous plants are again backcrossed to the recurrent parent. This is called the second backcross, and the progeny will once again consist of 50% heterozygous resistant plants and 50% homozygous susceptible plants. The backcrossing is thus continued for five to six generations, when the recurrent variety has been fully recovered along with the resistance gene that was contributed by the donor parent at the beginning of the crossing programme.

When the resistance gene being transferred is recessive, the breeder will have to grow the F_2 generation also to recover resistant plants for backcrossing to the recurrent parent. Obviously, the transfer of a recessive gene

Dominant gene for resistance (R)

	Donor parent	×	**Recurrent parent**
P_1:	RR		(Recipient) rr
F_1		Rr (Resistant)	
		↙	
BC_1	Rr	×	rr
BC_1F_1	1 Rr ↓	:	1 rr
BC_2	Rr	×	rr
BC_2F_1	1 Rr ↓	:	1 rr
BC_3	Rr	×	rr
BC_3F_1	1 Rr	:	1 rr

Continue backcrossing the resistant plants (Rr) with the susceptible variety in this way until BC_6 to $_7$, when the recurrent variety will have become resistant along with recovering almost its entire genotype. Test and finalize selection.

Figure 3.4 Incorporation of a dominant resistance gene (R) into a commercial susceptible variety by the backcross method.

into a commercial variety takes longer than that of a dominant gene. It is important to note that following a backcross, both F_1 and F_2 generations are grown. It is from the F_2 generation that the resistant plants are selected for further backcrossing. The backcrossing cycle may be continued for six to eight generations depending upon the need of the breeding programme and the resources in hand. The process for incorporating a recessive gene for resistance is schematized in Figure 3.5.

In both dominant and recessive gene incorporation, an evaluation of either the F_1 or F_2 from each backcross generation may need to be made under experimental conditions of disease or insect pressure, etc. in order to identify the correct genotype.

An advantage of the backcross method is that minimum yield testing is needed before release. The new variety usually keeps the old name but adds the year of release, e.g. 'Pirsabak–85'. A disadvantage of this method is that there is little opportunity for yield increase since the trait incorporated is yield-protecting rather than yield-enhancing.

Recessive gene for resistance (r)

	Donor parent		**Recessive parent**
P_1 :	rr	×	RR
F_1		Rr (susceptible)	
F_2		1 RR : 2 Rr : 1 rr (resistant) ↙	
BC_1	rr	×	RR
BC_1F_1		Rr (susceptible)	
BC_1F_2		1 RR : 2 Rr : 1 rr (resistant) ↙	
BC_2	rr	×	RR
BC_2F_1		Rr (susceptible)	
BC_2F_2		1 RR : 2 Rr : 1 rr (resistant) ↙	
BC_3	rr	×	RR
BC_3F_1		Rr (susceptible)	
BC_3F_2		1 RR : 2 Rr : 1 rr (resistant)	

Continue in this way until BC_{6-7}, when resistance has been incorporated into a fully recovered genotype of the recurrent parent (variety). Conduct necessary tests before release.

Figure 3.5 Incorporation of a recessive resistance gene (r) into a susceptible commercial variety by the backcross method

Sometimes the breeding material may have to be subjected to an appropriate mutagenic agent, particularly when it is intended to break undesirable linkages. When this has to be done, the time required to complete the whole operation will be longer and selection will have to be made very carefully.

STUDY QUESTIONS

1. What is a pure line? Can it also be developed in cross-fertilized species for direct commercial use?
2. What is mass selection? Why is it not widely used in the breeding of self-pollinated crops?
3. Give the defining characteristics of the pedigree method of selection, highlighting its advantages and disadvantages.
4. The bulk method is considered a simple and inexpensive method of breeding. Why is it not generally preferred over the pedigree method?
5. Discuss various modifications of the bulk method of breeding. Are they better than the original concept? If so, explain how they are better.
6. SSD is a relatively new method of breeding self-pollinators. How does it differ from the classical bulk method?
7. Explain the backcross method of breeding. Illustrate the mechanics of transfer of a dominant and recessive gene separately to a recurrent parent.
8. How many generations will it take to restore the genotype of the recurrent parent in a backcrossing programme? Explain in detail.
9. Explain how on selfing homozygosity increases and heterozygosity decreases.
10. In designing a hybridization programme, how would you decide on the choice of proper parents?
11. Why are self-pollinated crops homozygous and cross-pollinated crops heterozygous?

REFERENCES

Allard, R. W. Principles of plant breeding. New York: John Wiley and Sons; 1960.

Fehr, W. R. Principles of cultivar development. Vol. I, Theory and technique. New York: Macmillan Publishing Co.; 1987.

Frey, K. J. ed. Plant breeding II. Proceedings of 2nd Plant Breeding Symposium held at Iowa State University, 1979. Ames, Iowa: The Iowa State University Press; 1981.

Poehlman, J. M.; Borthakur, D. Breeding Asian field crops. New Delhi (India): Oxford and IBH Publishing Co.; 1972.

Poehlman, J. M. Breeding field crops. Westport, Connecticut: AVI Publishing Co.; 1987.

Rehman, A. Manual of wheat breeding procedures. Rome: Food and Agriculture Organization of the United Nations; 1987.

Simmonds, N. W. Principles of crop improvement. London: Longman Group, Ltd.; 1979.

Outline of Chapter 4

4. METHODS OF BREEDING CROSS-POLLINATED CROPS

Muhammad Saleem
Hidayat-ur-Rehman[1]

LEARNING OBJECTIVES

After working with this chapter, a student should be able to:

- Define heterosis, inbreeding depression, genetic variability, half-sib, full-sib, recurrent selection, and combining ability.
- Describe the techniques of various breeding methods by which new varieties of cross-pollinated crops are developed, and explain the advantages and disadvantages of each.
- Develop the following hybrids from four inbred lines A, B, C, and D of maize: single cross, double cross, modified cross, three-way cross, and backcross.
- Explain the differences between a synthetic variety and a hybrid of cross-pollinated crops, and be able to list at least two examples of each.
- Differentiate between different types of recurrent selection on the basis of the gene action involved or the testers used.

Breeding methods for cross-pollinated crops are not as well defined as those used for self-pollinated crops, because of the different population structures of cross-pollinated crops. In cross-fertilized or often cross-fertilized species, both self and cross-pollination may occur at the same time. Thus there are two forces operating: selfing, which leads to homozygosity, and out-crossing, which increases heterozygosity. For this reason, breeding methods will vary for each particular crop, depending on its mode of reproduction.

Plants of self-fertilized crops are homozygous; whereas, because of natural out-crossing, plants of cross-fertilized species are highly heterozygous. Therefore it is important for the breeder to understand the results

---

[1]Dr. Muhammad Saleem and Dr. Hidayat-ur-Rehman are with the Department of Plant Breeding and Genetics, NWFP Agricultural University, Peshawar.

of selection in a mixed population. Individual plant selections have often been used to develop uniform, pure-line varieties of self-fertilized crops. But in cross-fertilized species *individual plants* are seldom selected to establish a variety. This is because inbreeding leads to reduction in vigour and productivity. Instead, a *group* of similar looking plants or progeny are used to develop a variety in these crops.

The principal methods of breeding by which new varieties of cross-pollinated crops can be developed are: (1) **introduction**, (2) **selection**, (3) **development of synthetic varieties**, and (4) **hybridization**. Each of the above methods will be described in detail.

4.1 Introduction

Some economic plants like maize, tobacco, potato, and upland cotton were imported into Asia for cultivation from the Americas (Poehlman et al. 1969). The type of new material to be introduced into certain production areas generally depends on the needs of the plant explorers. Introduced material may be used in several different ways as described below.

1. As a source of new crops or of improved crop varieties. A considerable number of forest trees and field and horticultural crops have been introduced into Pakistan from time to time in the past. New forage crops and many ornamental plants have only recently been introduced from other countries and tested for adaptability and quality. Introduced stocks readily adapt to a new region of production and are grown commercially if soil and climatic conditions of the two countries are similar. Cross-pollinated species acclimatize more rapidly than self-pollinated crops due to frequent cross-pollination and gene recombination.

2. As a source of desirable types.

3. As a source of one or more desirable genes for incorporation into the local adapted varieties. Examples of such traits could be disease resistance, early maturity, cold tolerance, or quality.

Each of the above methods by which commercial varieties of field crops can be established is explained with examples as follows. An example of (1) is the introduction of a number of maize hybrids from the USA, Canada, and Australia into Pakistan in the 1950's and early 1960's. Some hybrids such as 'US 13', 'Indiana 909', and 'US 523' were fairly well adapted to the Pakistan environment and were high-yielding, but were not accepted by the farmers for the following reasons. (a) They were late-maturing and could not fit into the prevailing cropping pattern. (b) They possessed dented grains, while cultivators prefer white flint types. (c) 'US 13' hybrid and some other types were unacceptable due to their yellow grain colour. (d) F_1 seed production was a tedious job in those days and farmers were not willing to purchase fresh seed at premium cost each year.

A classic example of (1) above also involves a variety of maize. A cultivar, 'Pirsabak 7930', selected at the Cereal Crops Research Institute (CCRI), NWFP, was introduced and released as a national variety by three African countries, Zambia, Togo, and Benin (CIMMYT 1989). Many other national programmes are using this germ plasm in their breeding programmes (Gelaw 1986). Similarly, a number of open-pollinated maize varieties from Pakistan have been introduced into Afghanistan. These varieties combine features of early maturity, disease resistance, and high yield, which fulfill the requirements of Afghan growers in some provinces. However, many varieties of cross-pollinated crops have been introduced into Pakistan from countries including Mexico, Thailand, and India.

Contribution mode (2) is illustrated by the selection of desirable types from introduced stocks of sugar beet and maize at some experiment stations in this country. Good and early plants were selected from 'Pirsabak 7930' by the national maize programme and released under the name of 'Gauhar' for general cultivation in Islamabad District.

The excellent short-duration maize varieties, 'Kissan' and 'Azam', are an example of (3), the selection of desirable traits. These varieties contain some introduced tropical germ plasm from CIMMYT in Mexico, combined with locally adapted types and later developed by full-sib family selection. Improved and adapted varieties already in cultivation are usually superior to most introduced varieties of many crops. However, introduced material can serve as a source of desirable genes such as disease or insect resistance, cold or winter hardiness, stiff straw, or any other character of interest.

Collection and preservation of the germ plasm of local varieties of standard field and horticultural crops is very important for the future breeder. These as well as introduced materials must be preserved in viable condition for an indefinite period of time. Climatic factors such as temperature and relative humidity and the moisture content of the stored material greatly affect the viability of the germ plasm. These factors are closely interrelated. Special seed storage facilities will be needed for long-term preservation of germ plasm in regions with a hot and humid climate.

4.2 Selection

Selection can be defined as the process of isolating desirable genotypes from undesirable ones. It is the basis of all plant breeding procedures. Individual plants or their progeny need to be isolated and evaluated. Man is interested in isolation of the genotypes most suitable for human needs. He is not immediately concerned with *natural* selection, which operates in nature due to various pressures present under natural conditions where a population is growing. Maize or common wheat as we know it today cannot exist under natural conditions.

Selection can be carried out in a number of ways, depending on the mode of reproduction and the type of breeding method used. Selection tends to change gene frequencies and is thus responsible for creating new gene pools. A plant breeder selects plants according to his specific needs, for example higher grain yield, sugar or protein content, or early maturity. Since artificial selection tends to change gene frequencies in genotypes, changes in the genetic makeup of a new variety are expected. Since selection can only increase the frequency of genes already existing in the genotypes, it is essential that genes for the desired trait under selection be present in the source population. It should, however, be remembered that identification or isolation of a desirable genotype is a difficult job for the plant breeder.

Population improvement. Breeding methods of cross-pollinated crops are generally based on the principles of population improvement. To develop a variety of a cross-pollinated species, individual plant selection is generally not practised, because segregation and out-crossing make it difficult to maintain the original plant type in the succeeding generations. In population improvement schemes, except in mass selection where initial plant selection is based on visual evaluation called phenotypic selection, some sort of progeny evaluation is made. Most of the examples of the application of breeding procedures presented in this chapter pertain to maize, which is one of the most important cross-pollinated crops of Pakistan.

The presence of additive genetic variance for most agronomic characters in adapted and stable maize varieties has been emphasized by many workers (Lonnquist 1964; Moll and Robinson 1967). These and earlier studies created interest in variety improvement programmes. A number of population improvement schemes were developed and studied, and appropriate modifications suggested by several researchers to further improve their efficiency (Gardner 1963; Hallauer and Miranda–Filho 1981; Sprague and Eberhart 1977). Many population improvement procedures utilize additive genetic variance. The resultant population after selection is improved for the traits under consideration and also serves as a better source population for derivation of inbred lines.

The efficiency of population selection schemes depends on the presence of genetic variability, population size, presence of genetic linkages, kind of experimental design used, the type of traits or characters under consideration, selection intensity, and the magnitude of genotype-environment interaction.

Various population selection schemes involve the following phases: (a) Generation of a suitable number of families, (b) identification of genetically superior individuals or progenies, and (c) recombination of the selected progeny for the next cycle of selection.

Population improvement may be divided into two broad types: (a) intrapopulation improvement and (b) interpopulation improvement.

Intrapopulation selection schemes include the following procedures: (1) mass selection, (2) ear-to-row selection, (3) half-sib family selection, (4) full-sib family selection, (5) selfed progeny selection, (6) recurrent selection for combining ability, (7) test cross selection, (8) unit selection, (9) reciprocal unit selection, (10) multiple convergent unit selection.

4.2.1 Mass selection

Mass selection is defined as the selection of crop plants on the basis of their phenotypic superiority in the field and pooling their seed together without testing their general or specific combining ability. It is the earliest and the least expensive procedure for improving crop plants. Phenotypically desirable plants are selected, harvested, and composited for the next generation without progeny testing. In this scheme, only the female parent is selected and no control over the male is exercised. The purpose of mass selection is to concentrate favourable genes in a population. This method is effective for those characters which have high heritability and can be easily recognized, preferably before pollination, and used as the basis of selection. Good progress from mass selection has been reported for traits such as maturity, plant and ear height, disease resistance, size of ear, percentage of oil and protein, and prolificacy.

Several varieties of maize such as 'Shaheen', 'Sarhad Yellow', and 'Changez' have been developed by the Maize Programme in NWFP, using mass selection or one of its modifications. Mass selection requires one generation per cycle, whereas other systems require more than one. When mass selection is being done, plants should be spaced more widely than the planting distance recommended for commercial crops. Since an individual plant is selected, it is important that the plant should express its maximum genetic potential.

Though progress from the mass selection and ear-to-row procedures has been reported by some researchers, the general opinion is that these two breeding procedures are generally ineffective for improving yield in adapted varieties. Some of the factors that minimize progress from mass selection are: (1) the effect of environment on the phenotype; (2) lack of parental control; and (3) low selection intensity, which leads to inbreeding depression.

Modified mass selection. Several modifications of mass selection have been suggested to overcome some of these limitations. To reduce the effects of environment on the phenotype, Gardner (1961) suggested a procedure called grid or stratified mass selection. He studied subpopulations of a maize variety in 40 plants each. This provided an opportunity to equally sample the environment of the experimental area. In four years, he was able to increase maize yield from 79.3 to 97.4 bu/acre (4988 to 6126 kg/ha). More recently, a modification of mass selection has been suggested by Burton and Gates (1991 personal communication) involving recurrent restricted phenotypic selection (RRPS). Mass selection is modified by a number of restrictions that make RRPS more efficient. Selection is restricted to control both male and female parents and thus doubles the efficiency of mass selection. Other restrictions include the precision of the selection process and improvement in random mating, and reduce the time required to complete a cycle from three years to one year. Burton and Gates report an increase of 11% in forage yield per cycle in bahiagrass, and also a per cycle annual increase of 11% in the forage yield of pearl millet through cycle six.

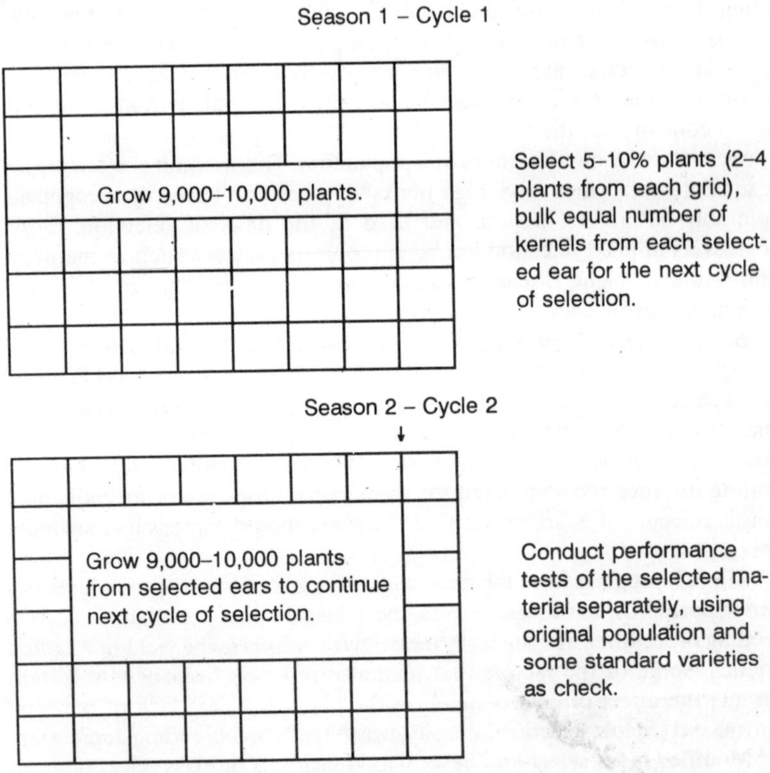

Figure 4.1 Illustration of modified mass selection. Procedure after Gardner (1961).

A step-by-step explanation of Gardner's modified mass selection procedure is outlined here. This procedure is illustrated in Figure 4.1.

a. Grow 9,000–10,000 plants in isolation.

b. Space plants widely apart (40,000 plants/ha) for full expression of the genetic potential of each individual plant.

c. Subdivide the field into several grids or subplots with 40–60 plants in each grid. The number of plants depends on the selection intensity to be used (5–10%). Select three or four plants in each grid.

d. Dry all ears to a uniform moisture content before weighing.

e. Bulk equal numbers of kernels from each selected plant to form a balanced composite. Prepare three sets of seeds: one for continuation of the next cycle of selection, the second for yield evaluation, and the third as remnant seed for possible failure of the first two sets caused by environmental factors.

One undesirable consequence of this method was the reported later maturity of the selected population. The apparent improvement in yield observed was thus partly contributed by late-maturing genotypes.

4.2.2 Ear-to-row selection

Ear-to-row selection involves the selection of a number of ears on the basis of some criteria, e.g. yield, lodging, maturity, or chemical composition. All ears are evaluated by a progeny test. Ears for the next cycle of selection are chosen from those progeny exhibiting the highest means for the traits under consideration. This method as originally employed was a form of mass selection, and not ear-to-row breeding as used today in half-sib progeny tests.

Many workers reported success from ear-to-row selection for characters such as plant and ear height, leaf area, and maturity. The system was later extended to grain yield. Failure to obtain continuing yield improvement in the past was attributed to inadequate field plot techniques. Reasonable rates of progress in yield improvement were achieved after modifications suggested by Lonnquist (1964) were adopted. In the modified ear-to-row selection scheme, the limitations of inbreeding and the use of single plots were minimized. The essential features of the modified ear-to-row system of breeding are given below.

a. Select a number of ears (250) from a genetically broad-based source population such as an open-pollinated variety, composite, synthetic variety, or advanced generations of successful hybrids.

b. Grow seed from each selected ear in a 16 × 16 triple lattice design using six standard varieties and hybrids as checks. Each of the three replications is grown in a different environment.

c. Set one trial in an isolated area for evaluation and progeny recombination, using balanced composite as a male parent, i.e. yield trial cum crossing block. Rows are planted in a ratio of four female to two male. Seeds for the male rows are obtained from a balanced composite of 250 ears. All female rows and check entries are detasseled, and pollen allowed from male rows in this replication.

d. Select the best 20% of the rows (i.e. 50 rows) on the basis of yield and other agronomic characteristics across three environments. The five best ears from each of these selected rows (a total of 250 ears) are retained to be used in the next cycle of selection. The original population should be maintained in a viable condition and without any selection, as all comparisons of the selected material will be made with this source population.

Ear-to-row selection is more effective than mass selection in that progeny are evaluated across locations, and selection is practised between and within progeny, which is not possible with mass selection. One cycle of selection is completed in one year and selection for more than one character can be simultaneously practised. Modified ear-to-row selection appears to be superior to half-sib selection because modified ear-to-row selection provides an opportunity to select among progeny as well as within a progeny group. Half-sib selection, on the other hand, is practised among progeny only (Sprague and Eberhart 1977). This scheme also allows selection of individuals based upon phenotypic performance (mass selection) in natural-crossing block isolation.

Webel and Lonnquist (1967) reported 9.4% yield gain per cycle from four cycles of modified ear-to-row selection. Gains over 10 cycles from the same study were 4.0-4.5% per cycle (Bahadur 1974), but undesirable attributes such as number of broken stalks and grain moisture percentage also increased.

4.2.3 Half-sib family selection

A half-sib family is defined as the progeny from a cross within a population having one parent in common. **Half-sib (HS) family selection** is similar to modified ear-to-row selection, but the population for the next cycle of HS selection is synthesized from the remnant seed of the selected progeny rather than from the bulk seed from the natural crossing block for the modified ear-to-row selection (Sprague and Eberhart 1977). Superior progeny are identified on the basis of the characteristics under consideration and are recombined in a natural recombination block in isolation. Recombination is usually accomplished in the off-season. In tropical or subtropical areas, at least two crops per year can be grown successfully. It is possible to complete one cycle of half-sib family selection in two seasons or one year. Breeding objectives, characters under improvement, the type of genetic material, and limitations imposed by the quantity of seed of the selected progeny available determine the number of test environments for half-sib family selection. 'Pahari', a recently released maize variety for mid-elevation zones, was developed and selected through the half-sib family selection procedure. This breeding procedure has been more extensively used by plant breeders than any other population improvement scheme.

Several modifications of this method are available, but the one used by maize breeders in Pakistan is outlined below and illustrated in Figure 4.2.

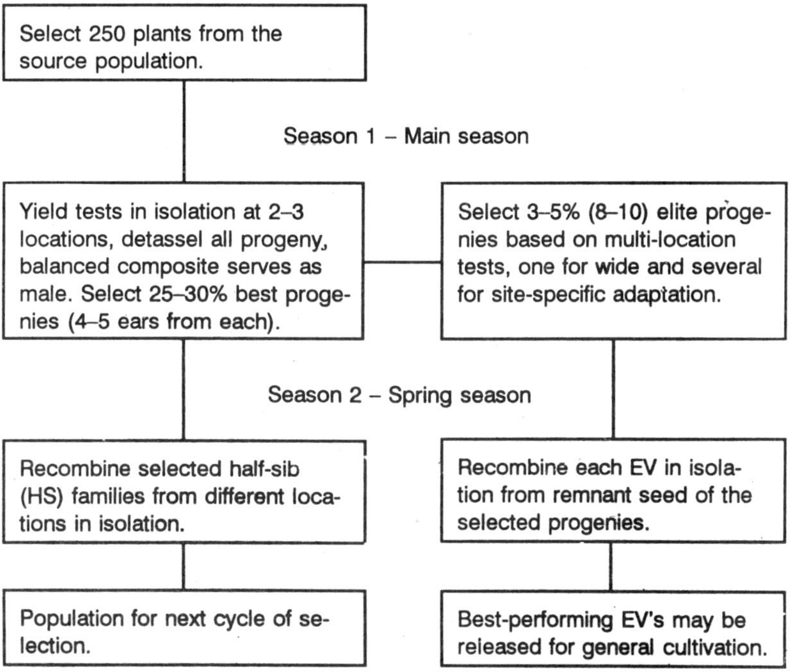

Figure 4.2 Half-sib family selection procedure as used in maize.

a. Grow about 250 selected half-sib progeny from a genetically broad-based population and arrange in a 16 × 16 simple lattice design. Use six standard varieties or hybrids of similar maturity group in each trial, at each location.

b. Grow the trial in isolation, preferably at more than two locations. Each progeny is grown in a one-row plot of 5 m length using two replications. Bulk an equal number of kernels from each progeny to obtain seed for the male rows. Usually four female to one or two male rows are used, depending on the pollen-producing capacity of the parental population and environmental conditions. To avoid contamination, detassel all check varieties and all female progenies before pollen shedding.

c. One of the replications is inoculated with important diseases of the production region to identify disease-resistant progeny. Alternatively, a companion nursery is grown separately and all progeny are inoculated.

d. Select progenies (about 40%) at several stages before harvest based on their phenotypic appearance and freedom from diseases. Retain one of two of the best ears from only 50% of the initially selected progenies at harvest. Selected progenies from all locations are grown as female rows for recombination in the natural crossing block. Seed from the 50–60% best plants is used as bulk male. All progenies as well as poor male plants are detasseled before pollen shedding. Pollen from desirable males is allowed to pollinate the progenies.

e. If necessary, some promising and matching material from advanced breeding programmes may be used as female rows in the recombination block to broaden the genetic base of the selected population. The performance of new material is tested as test crosses. The best progeny from the introduced material may or may not be subsequently merged with the main population, depending on their acceptability.

f. One cycle of this selection scheme is completed in two seasons or one year, since recombination and generation of the progeny is accomplished simultaneously in the crossing block. Four or five ears from each selected progeny are retained. These are tested in yield trials to continue the next cycle of selection.

g. About 8–10 outstanding progeny from the ones selected across locations are synthesized at the same time to give rise to an improved experimental variety for wide adaptation and two or three experimental varieties for site-specific adaptation. The new versions are also included in the performance tests as some of the checks. The experimental varieties may be released as commercial varieties if proved superior in yield trials. Seed is increased and distributed for general cultivation. The new version of 'Changez', a short-duration maize variety, was developed in a similar manner.

4.2.4 Full-sib family selection

While half-sib progeny have only one parent in common, **full-sib selection** utilizes crosses in which the progeny have both parents in common. Real brothers and sisters exemplify genetically full-sibs in a family. This selection procedure is considered twice as effective as half-sib selection because it provides control over both parents. Full-sib progeny are evaluated for yield and other agronomic traits in replicated yield trials across locations. The selected progeny are recombined from remnant seed for one or two generations in isolation before a new set of full-sib progeny are generated. The number of generations required for random mating depends on the objectives of the breeding programme.

Full-sib selection requires two generations or one year per cycle in regions where two crop seasons are available. In this case recombination and formation of families are accomplished simultaneously in the first season, and yield tests in the second (Sprague and Eberhart 1977). It requires three seasons where recombination and formation of families are carried out in separate seasons. A step-by-step procedure for full-sib family selection with some modifications is detailed below and illustrated in Figure 4.3.

1. Full-sib crosses are generated from a high-yielding base population in the spring or off-season, while evaluation of the progeny is made in the kharif or main season. Usually 300–400 plant-to-plant reciprocal crosses are made among selected progeny and between good plants. At harvest, undesirable crosses are discarded and reciprocal ears from each cross are bulked to provide more seed for testing in different environments. Full-sib crosses between matching plants are made so that they are uniform for maturity, plant and ear height, and other agronomic characters. This is an important consideration because a commercial variety will eventually be developed from 8-10 elite, full-sib families, which should give rise to an agronomically acceptable cultivar. At harvest, 250 progeny are finally retained.

2. Progeny are evaluated in a 16 × 16 simple lattice design in the main season at three or four locations. A set of six standard checks is also included in the performance trial. In addition to yield, all progeny are evaluated extensively for resistance to diseases, insect pests, or any other problem of the production area. Artificial inoculation for important diseases is carried out if necessary. Remnant seed of all progeny is reserved for subsequent use of the 25–30% selected progeny to form the succeeding cycle population and to constitute improved versions (for

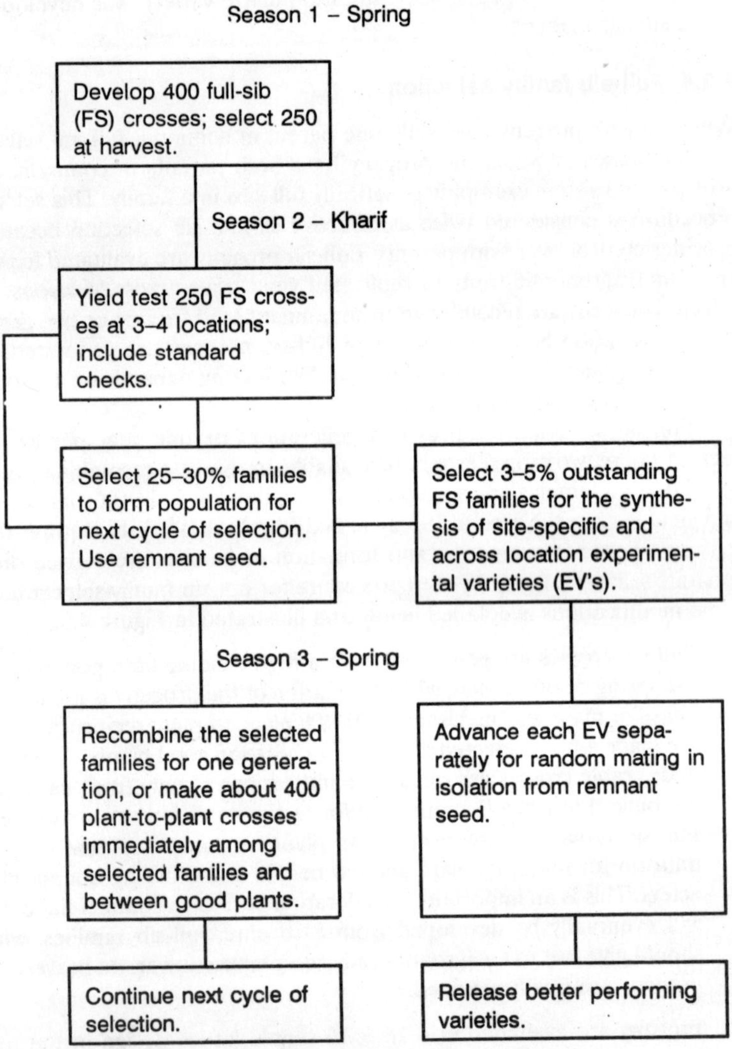

Figure 4.3 Steps of full-sib family selection procedure

wide and specific adaptation) of the population from 3–5% outstanding families.

3. Selected (60–75) full-sib progeny are recombined to generate a population for the next cycle of selection and 8–10 elite progenies to generate improved varieties. Recombination is accomplished in the off-season. Improved experimental varieties (EV's) may be developed as site-specific or across locations from the 8–10 best performing progenies. Each EV is recombined by **chain crossing**, a procedure in which individual genotypes in a series are mated with each other. For example, in the set of genotypes {A, B, C, D}, the pollen (male) from A is applied to the silk (female) of B; the pollen of B to the silk of C; the pollen of C to the silk of D; and the pollen of D to the silk of A, thus completing the chain. An equal number of seeds from each chain cross is bulked for further random mating and for inclusion in replicated performance tests against the parental source population and some standard commercial varieties.

Variable yield gains from full-sib family selection have been reported by many workers, but yield increase usually ranged from 3 to 4% per cycle (Singh 1983). Yield increase of 7.5% per cycle through five cycles has been reported, although progress in subsequent cycles was reduced to 3.0%. Paliwal and Sprague (1981), improving different maize populations for the developing world, observed yield increase of 0.75–9.8% per cycle from this selection scheme. Moll and Stuber (1971) observed 2.5–4% increase in yield per cycle in five maize varieties. Many maize varieties have been developed in this country using full-sib family selection. 'Kissan', a white, short-duration variety, was developed through full-sib family selection

With each cycle of full-sib family selection, one EV for general adaptation and several site-specific EV's are generated. It is observed that improved varieties developed for wide adaptation (across locations) are generally superior in yield and adaptability than improved varieties developed for specific locations.

4.2.5 Selfed progeny selection

Selfed progeny or S_1 and S_2 progeny selection is considered an efficient method of plant breeding but has not been utilized or tested as widely as other population improvement methods. S_1 and S_2 progeny selection is used to improve the performance of elite inbred lines and/or that of populations with a broad genetic base, depending on the objectives of the breeding programme. This scheme is used to identify superior progenies, which are recombined from the remnant selfed seed of the selected progenies for the next cycle. The advantages of this scheme are reduction in inbreeding depression (increased genetic variability among the progenies) and the elimination of lethal and sublethal recessive genes from the population (Hallauer and Miranda–Filho 1981). The drawbacks include longer cycle duration, greater experimental error, and linkage. However, some or all of these weaknesses can be avoided if progenies are handled carefully. This breeding scheme holds considerable promise as a method of population improvement in the future.

S_1 progeny selection requires four crop seasons to complete a cycle of selection for characters which are quantitative in nature and cannot be identified before flowering. However, only one year or two crop seasons are required for traits which can be identified before pollination. In this case evaluation and recombination can be carried out simultaneously.

In S_2 progeny selection, S_1 progeny are evaluated in the kharif season for characters such as disease or insect resistance, and the selected S_1's are advanced to S_2 by making one or two pollinations in each selected S_1 in the off-season nursery. While S_2 progeny are tested for yield and other traits in the main season, approximately 20–25 outstanding progeny, using 10% selection intensity, are recombined to generate the first cycle of selection. Random mating is necessary before the generation of another cycle of improvement.

Many people have reported improvement by using this selection procedure. Some reported improved resistance to stalk rot; others observed resistance to leaf feeding by maize borers from this selection scheme. This procedure is considered effective in improving yield and other agronomic traits.

4.2.6 Recurrent selection

Recurrent selection (also called **cumulative selection**) is a breeding scheme to concentrate favourable genes for yield or other characteristics by repeated cycles of selection, and to increase genetic recombination in a population. The objective of recurrent selection is to increase the frequency of desirable genes without marked loss of genetic variability in a population, i.e. to increase favourable genes while also increasing chances for genetic recombination.

General outline of recurrent selection in cross-pollinated crops (maize).

1. Plants from a mixed population are selected.
2. Plants are simultaneously selfed and evaluated (out-crossed to a tester) for characters under improvement.
3. Plants with inferior performance for characters are discarded.
4. Plants with superior performance are propagated from the selfed seed.
5. Plants in (4) are crossed in all possible combinations by hand or in isolation.

6. The intercrosses from (5) serve as source population for another cycle of selection and intercrossing.

7. The cycle may be repeated as long as selection continues to yield progress. Choice of source population can be made from open-pollinated varieties, a synthetic variety, progeny of intercrosses of inbred lines, or single or double-cross hybrids, depending on breeding objectives.

Final use of populations developed by recurrent selection can be made in various ways: (a) for the production of inbred lines; (b) for crossing with inbred lines, single crosses or other improved populations to form hybrid varieties; (c) as source foundation stocks for synthetic varieties.

Four types of recurrent selection are recognized. These differ in the assumptions with respect to the importance of gene action, and in the use of a tester.

a. Simple recurrent selection. This is the simplest form of recurrent selection. Plants are selected or discarded on the basis of their phenotypes or their S_1 progeny. Since no progeny test is required, the use of this procedure is limited to highly heritable characters. Each cycle may be as short as one season if evaluation is possible before pollination. Jenkins et al. (1954) developed leaf blight resistance in maize using simple recurrent selection. The stepwise procedure is outlined below.

1. Grow a source population and select a number of individual plants based on superior phenotype.

2. Grow progeny and cross in all possible combinations. Grow a new population of crossed seed. Repeat cycle as long as progress can be realized.

b. Recurrent selection for general combining ability. Plants from a heterozygous source population are selected, selfed, and out-crossed to a heterozygous tester to identify the plants with good combining ability. In maize, selfing and out-crossing are possible in two ways: either using the pollen for out-crossing, or using a second ear on the same plant to receive out-cross pollen. The selected plants are propagated from selfed seed and intercrossed in all possible combinations; then a composite of the intercrossed seed is used as a source population for the next cycle. This process primarily emphasizes additive gene effects. Its simplest form takes three generations per cycle. This method is the direct outgrowth of experiments on early testing (Jenkins et al. 1940). Evidence of its usefulness comes from studies on maize (Lonnquist and McGill 1955). The procedure of recurrent selection for general combining ability is outlined as follows.

1. Self-pollinate several plants, out-cross to a tester with a broad genetic base. Save selfed seed.

2. Plant test crosses and evaluate. Select plants with superior general combining ability.

3. Grow selfed seed of superior plants, and intercross them.

4. Repeat cycle as long as desired.

c. Recurrent selection for specific combining ability. Hull (1945) proposed recurrent selection for specific combining ability. He believed that over-dominance was the major type of gene action involved. The pattern of selection is the same as that for recurrent selection for general combining ability, except that a tester which has a narrow genetic base, i.e. an inbred line or a single cross, is used. Progress from this method has been reported by Sprague et al. (1959) in two open-pollinated varieties of maize.

Interpopulation selection includes improvement methods like reciprocal recurrent selection, half-sib reciprocal selection, and full-sib reciprocal selection. Reciprocal recurrent selection will be discussed briefly. There are several modifications which further refine many of these methods, but these will not be discussed in this chapter.

d. Reciprocal recurrent selection. Comstock et al. (1949) proposed a procedure that would be useful in selecting simultaneously for both specific and general combining ability. Two genetically broad-based populations (A and B) are selected. This method varies from recurrent selection for general combining ability in that population A is used as tester for B and B as tester for A. With each cycle of selection the two newly synthesized improved versions of A and B (A′ and B′) are used as testers. The two source populations used in reciprocal recurrent selection (RRS) should be genetically unrelated and have good combining ability. Two synthetic varieties, two open-pollinated varieties, or two single crosses may be used as source populations. Sufficient plants (400–500) should be selected to avoid inbreeding as much as possible. The procedure of reciprocal recurrent selection is outlined as follows

1. Grow two genetically different populations. In each population select and self several superior plants and simultaneously cross them to the other population. Save the selfs in reserve.

2. Grow and evaluate the test crosses. Identify the original individuals with superior combining ability.

3. Grow the reserve selfed seed of each superior original plant. Keep each population separate and intercross within each population. This completes one cycle of RRS. Repeat the cycle for as long as progress can be made.

Many modifications of recurrent selection are now available. All breeding methods which involve selection of superior progenies and their combination to constitute a population through cyclic selection can be included in recurrent selection. Recurrent selection procedures have been successfully used in cross-pollinated crops; however, it is also being utilized in self-fertilized crops where suitable male sterility genes are available.

4.3 Synthetic varieties

A synthetic variety is a multiple hybrid between selected genotypes which combine well among themselves; after it is initially constituted, it is maintained by random crossing. It is developed from component parts with known general combining ability. This differentiates synthetic varieties from those developed by mass selection or line breeding. The genetic base of a new synthetic variety may vary from a few selected plants to a mixture of several genetically different lines, clones, or varieties. This depends very much on the intended use of such varieties in areas with varying production capabilities, i.e. the technological level of the farming community. A synthetic variety may be reconstituted from time to time from the component units, which should be maintained in the breeding programme in a viable condition.

The objective of testing in the development of synthetic varieties is to identify genotypes that combine well among themselves. Testing methods vary from simple visual identification of highly heritable characters to progeny tests for traits like yield and quality.

In countries where the annual production of first-generation seed is not practical, synthetic varieties offer an opportunity to take advantage of heterosis. The use of synthetic varieties has been suggested for the improvement of crops like forage varieties, sugar beet, maize, sunflower, and other cross-pollinated crops. Synthetic varieties offer a good opportunity to exploit controlled utilization of heterosis in forage crops and maize. In Pakistan and several other countries, a number of synthetic varieties of various maturity groups have been developed which favourably compare with the best F_1 maize hybrids of similar maturity.

Advantages of synthetic varieties.

1. Synthetic varieties are of considerable importance in countries where the seed industry is not developed. In areas where hybrid seed production and distribution is not feasible because of the subsistence level of the farmers, growers can save their seed for the next crop instead of purchasing fresh seed each year at premium prices.

2. Synthetic varieties can be of considerable value in marginal areas of production.
3. The greater genetic variability of synthetic varieties permits more flexibility and adaptability to environmental stresses.
4. Improved synthetic varieties can be used as sources of favourable alleles for derivation of inbred lines.
5. Development of synthetic varieties is relatively simple and takes less time than that of hybrids.

Factors that may affect the performance of a synthetic variety for a particular area are: (a) choice of source material and number of parents involved; (b) procedure used for the synthesis of a synthetic variety; (c) maintenance of the synthetic variety in later generations.

Synthetic varieties have been successfully used for the improvement of cross-pollinated forage crops. The population improvement schemes such as half-sib, full-sib, or S_1 selection described earlier in this chapter cannot be used in the development of synthetic varieties of forage crops for the following reasons. (a) Progeny performance of individual plants in most forage crops cannot be determined. (b) Crossing in many forage crops is tedious because they have very small floral parts. (c) Many forage species are inbreeding intolerant; thus there are some forage crops for which certain breeding procedures cannot be used. Some species of grasses including napiergrass and elephant grass and forage legumes like lucerne, red clover, and white clover are examples of such types. However, synthetic varieties of forage crops, maize, sugar beet, sunflower, and other cross-pollinated crops have been successfully produced.

The procedure for the constitution of a synthetic variety has the following features.

1. A synthetic variety is developed from reproducible component parts.
2. The component units of the synthetic are selected on the basis of their ability to combine among themselves.
3. The components are preferably inter-pollinated either through chain crossing or diallel crosses, a procedure in which a set of genotypes is crossed in all possible combinations.
4. The component units are maintained separately so that the synthetic variety can be reconstituted at regular intervals.

The procedure for the development of a synthetic forage crop, which involves several steps, is illustrated in Figure 4.4.

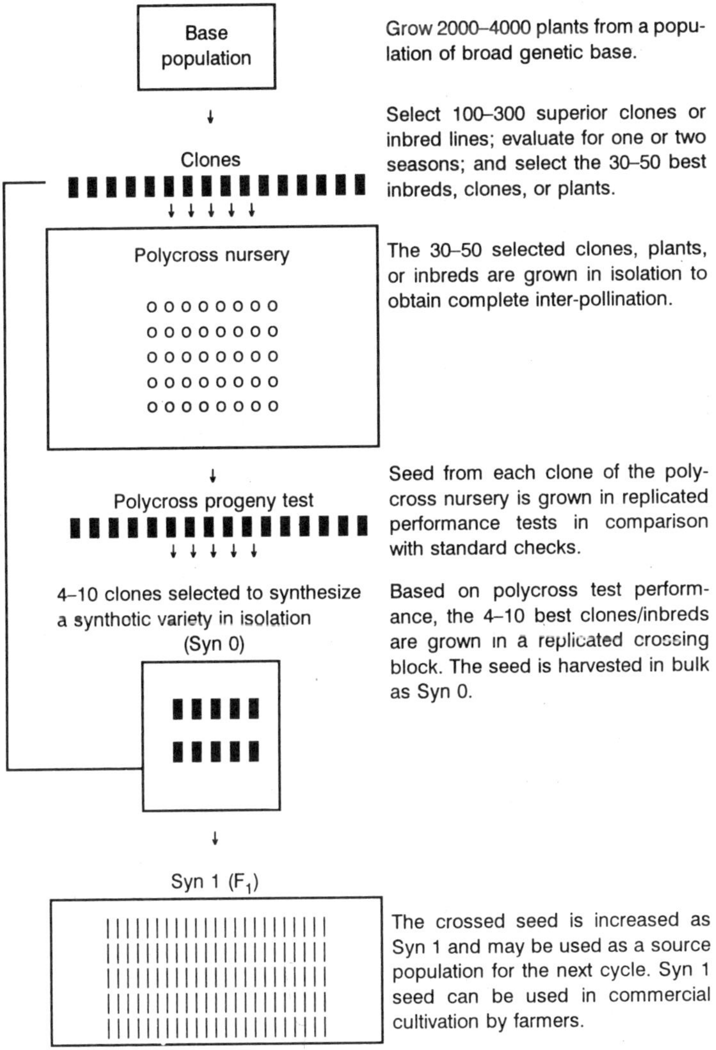

Figure 4.4 Procedure for development of a synthetic forage variety using a polycross test (adapted from Poehlman 1987).

A population having a broad genetic base should be used as source population. A source nursery of the vigorous and productive clones is established by selecting several hundred superior plants using mass selection or any other breeding procedure. The plants are selected keeping in view the specific needs of the production region. Thirty to 50 phenotypically superior plants are selected for the establishment of a polycross nursery and to facilitate random crossing. Several replications are used to allow random pollination more completely in the polycross nursery. The seed from each clone is harvested and kept separately as polycrossed seed.

The polycrossed seed of each clone is grown in replicated tests in several locations for evaluation of yield and other characteristics. The test includes one or two standard check entries and is grown under conditions similar to those used by the farming community.

The appropriate number of lines that should constitute a synthetic variety is a matter of debate. Generally, four to 10 clones are suggested to be a reasonable number for a synthetic forage variety. Use of fewer lines may result in a narrow genetic base and a narrow range of adaptation. The component clones are established in a crossing block to allow random mating of all components. This constitutes the **Syn 0** generation of the synthetic.For random mating, it is important that all component units flower at the same time. If the units do not flower at the same time, early-flowering plants will cross with other early-flowering plants and late-flowering with late-flowering. This will lead to the formation of sub-populations in the synthetic variety, which will be highly variable for several characters.

The next generation is called **Syn 1** (equivalent to the F_1 of a hybrid); it may exhibit heterosis if inbred lines were used to synthesize the variety. The yield of the Syn 2 (F_2) generation is normally somewhat reduced because of production of some homozygous genotypes, which lead to inbreeding depression and reduce vigour. The synthetic variety can be better evaluated from the performance of the Syn 2 generation in replicated trials over a number of environments.

Synthetic varieties of maize with a broad genetic base have been developed by CIMMYT (International Maize and Wheat Improvement Centre) in Mexico for developing countries throughout the world. The national programmes then can select germ plasm for disease resistance, adaptation, maturity, drought or cold tolerance, yield, or any other trait according to their local needs.

4.4 Hybridization

Hybridization under natural conditions is a powerful factor in the evolution of plants. It is also a powerful tool in the hands of the plant breeder for the development of specific cultivars for man's use. As early as 1761–66, Koelreuter observed hybrid vigour in a number of F_1 crosses. But it was East (1908) and Shull (1909) whose experiments in maize led to the interpretation of inbreeding depression and the restoration of vigour upon crossing.

This early work and the suggestion of double-cross hybridization by Jones (1918) for the general crop still form the basis of modern maize breeding. The same principles have been applied equally well to many other cross-pollinated crops.

Hybrid maize is the first-generation progeny of a cross between genetically unrelated parents, usually inbred lines. Producing hybrid maize involves the following steps: (a) Development of vigorous inbred lines with high grain yield; (b) *per se* evaluation of the inbred lines and in hybrid combination; (c) commercial production of the hybrids; and (d) further improvement of the lines for yield, and for disease and insect resistance.

4.4.1 Development of inbred lines

An inbred line is a pure line developed by self-fertilization and selection until homozygous plants are obtained. With naturally cross-pollinated crops, artificial control of pollination is necessary to develop inbred lines. Maize can be used as an example. Developing ear shoots are covered with a shoot bag before silk emergence. Tassels are bagged one day before pollination, and pollen grains from within the bag are transferred to the silks of the same plant the next day. Care must be taken to avoid contamination from foreign pollen. Usually five to seven generations of inbreeding are necessary to develop true-breeding lines.

Inbred lines can be developed from open-pollinated varieties, synthetics, composites, or advanced generations of hybrids. Choice of the base population from which inbreds are derived is important, since the best inbred lines will be derived from the best-performing source populations.

Inbreeding should be initiated simultaneously in two genetically diverse populations known for their high combining ability with each other. Inbreeding is often accompanied by selection for attributes such as vigour, disease resistance, standability, maturity, or any other desirable character. Weak and undesirable plants are discarded at harvest. Ears from the best plants are saved and planted ear-to-row the following season. Selection is practised among and between progeny rows. Only superior lines with desirable attributes are used for further inbreeding.

Selection during line development identifies vigorous plants with potential commercial value and can help eliminate lines with no commercial usefulness. Visual selection during inbreeding may lead to improved hybrid performance. Some workers have reported that selection during inbreeding has little direct effect on combining ability for yield of an inbred line. Selection, however, is effective in modifying lines for general vigour, maturity, and disease and insect resistance, which are important in the production of commercial hybrids. After four or five generations of selfing the lines are separated into distinct and relatively true-breeding types.

Inbred line development is easy, but to identify lines which combine well is rather difficult. The final merit of any line is judged by its performance in hybrid combination. The success of a commercial hybrid may depend upon the potential of the parental inbred lines, particularly the female parent. Conventional hybrids, especially single crosses, are specific to certain environments. On the other hand, hybrids developed from short-term inbred lines, particularly in tropical or subtropical regions, are more useful. These hybrids possess relatively more variability, are more productive, and can withstand unfavourable environments better than hybrids of long-term inbred lines.

Other methods of pure line development include the single hill method, pedigree selection, and the production of homozygous diploids.

Inbreeding depression. The effects of inbreeding in normally cross-pollinated crops have been known to plant breeders for a long time. Inbreeding generally is accompanied by appearance of a large number of lethal or sublethal types in the early generations. Hallauer and Sears (1973) reported inbreeding depression to be the result of accumulation of homozygous recessive undesirable loci. However, some useful types such as resistance to smut and leaf blight may also appear. Other effects of inbreeding in cross-pollinated crops include increase in homozygosity, production of uniform lines, and decline in vigour and productivity. Inbreeding effects are more severe during the first few generations of selfing. Jones (1939) observed that there was no appreciable reduction in plant height after five generations while yield reduction ceased after 20 generations of self-fertilization.

Loss in vigour and accompanying reduction in productivity following inbreeding in most cross-fertilized crops is of great concern to plant breeders. Rapid decrease in yield, plant height, ear length, and other traits, particularly in the early years of self-fertilization, and the restoration of vigour and productivity upon crossing unrelated genotypes have posed great difficulty in explaining the genetic basis of inbreeding depression.

Some cross-pollinated crops are more tolerant of inbreeding than others. Crops like cucurbits, onion, sunflower, rye, maize, and some kinds of grasses are more tolerant of inbreeding than alfalfa and carrot, which hardly can be maintained beyond the S_3 generation.

Heterosis or hybrid vigour. Heterosis or hybrid vigour is defined as the increase in size and vigour of a hybrid over the best parent. If there is no such increase, there is no heterosis. By definition, heterosis is over-dominance and is synonymous with the term *hybrid vigour*.

Utilization of heterosis is widespread in plants grown for economic return (e.g. maize, sorghum, tomato, onion); insects (silkworm); and livestock (poultry, beef cattle). The plant breeder is generally interested in higher grain yield or increased vegetative growth, but heterosis may result in greater cell size, plant height, leaf area, root development, ear or grain size, grain number, and so on. One of the most fascinating and productive examples of the utilization of heterosis has been the commercial production of hybrid maize.

The genetic basis of heterosis is not completely understood, and many possible explanations of heterosis have been offered. Several theories have been proposed under the headings of (a) simple gene action, (b) heterozygosity *per se*, (c) dominant and partially dominant growth factors, (d) physiological aspects, (e) multiple alleles, (f) over- or super-dominance, (g) cytoplasm, and (h) additive effects. But none of these explanations appears adequate to cover all cases of heterosis.

It is believed that heterosis is due to heterozygosity, which increases the stimulus to development, which is less when similar genes are received from both parents. Heterozygosity *per se* is not an explanation of heterosis, but it is the state of conditions when heterosis is expressed. That is, individuals which display heterosis are heterozygous, but not all heterozygous individuals show heterosis.

East (1936) later extended the theory to include a series of alleles, a, a_1, a_2, a_3, a_4, . . . at a single locus. In this hypothesis, heterosis increases with greater differences in the combined alleles in a series $a_1a_2 < a_1a_3 < a_1a_4$, etc. This would provide varying amounts of over-dominance depending on the presence of certain alleles.

Another explanation is the accumulation of favourable dominant genes in the F_1, some of which are contributed by one parent and some by the other. When the F_1 hybrid contains dominant genes at many loci, it would exhibit more vigour than either of the parents. If this theory is correct, it should be possible to concentrate sufficient favourable dominant genes in a homozygous inbred line which is as productive as the parent line (open-pollinated variety from which the inbred line is derived). However, since such homozygous individuals equalling the open-pollinated varieties in vigour have not been observed, the theory is criticized on this basis. Heterosis is a complex phenomenon and cannot be explained on the basis of a single reaction.

4.4.2 Evaluation of inbred lines

***Per se* evaluation of inbred lines.** The yield potential of the inbred lines of concern is visually evaluated. Characters like yield *per se*, ear characteristics, pollen-shedding ability, resistance to diseases, and vigour of the lines are considered. Poor lines are discarded.

Evaluation of inbred lines in hybrid combinations. The combining ability of inbred lines in hybrid combinations and their *per se* performance (yield and other agronomic traits) may be simultaneously evaluated to save time and money. Early testing at the S_1 level is effective in eliminating a large number of poor combiners in the early generations of inbreeding. This helps the breeder to study only promising lines in the later generations. Lines are initially evaluated for general combining ability and the selected lines are subsequently tested in hybrid combinations to determine their specific combining ability.

i. *General combining ability.* General combining ability (GCA) of an inbred line is determined in top crosses. A locally adapted open-pollinated variety may be used as a tester to classify the lines on the basis of general combining ability. It is desirable to use one or more testers with recessive alleles at loci for which selection is practised.

Top crosses are produced by growing inbreds as female in the ratio of four female to two male rows. Single-row plots of 5 m length are grown to get enough seed for multi-location testing in replicated trials. All inbred lines are detasseled before pollen is shed. The male is usually split-planted at two different dates to ensure pollen supply over a longer period of time. Inbreds are visually evaluated in the top-cross production block for agronomic traits and yield potential. Undesirable ones are discarded.

Top crosses are evaluated in replicated yield trials in several locations. These crosses are arranged into a number of 5 × 5 or 7 × 7 simple lattice trials for greater experimental precision. Two or more standard checks, such as standard commercial hybrids including a tester parent, are included in each trial. Better performing lines with good agronomic characters are retained for further evaluation.

ii. *Specific combining ability.* Lines selected for superior general combining ability should be divided into groups of 10 each to make crosses in all possible combinations. Inbred lines in each group should be genetically unrelated and of similar maturity with a good level of disease resistance.

All possible crosses are made among the selected lines. Reciprocal crosses are made to have more seed for subsequent testing in several locations. All single-cross hybrids are evaluated in replicated performance tests in from four to six environments. Standard commercial hybrids are included as check entries. Based on data from hybrid performance, the specific combining ability of the parent lines is determined.

4.4.3 Commercial production of hybrids

The selected inbred lines which nick (combine) well can be combined into various types of hybrids which differ in the number of lines involved and the specific manner in which these are combined. Combinations of inbred lines that produce superior hybrids are generally rare. Most inbred lines are discarded in the testing programme, and only a few are extensively used in commercial hybrid seed production. Various types of hybrids are described in the following paragraphs.

i. Single-cross hybrids. A single cross (A × B) is the hybrid progeny of a cross involving two genetically unrelated inbred lines. Single crosses are highly heterozygous and are more vigorous and productive than the original open-pollinated variety from which these lines were derived. The commercial use of single-cross hybrids has substantially increased in many countries. The use of modified single-cross (A × A') × B hybrids has facilitated the commercial production of hybrid seed at relatively lower cost. The modified single cross utilizes a cross of two lines selected from the same inbred population which differ slightly but sufficiently to give some hybrid vigour and thus increase the seed yield when used as the female. An example of a good single-cross maize is AUH 101, which was developed from two inbred lines AU 292 and AU 302.

ii. Three-way hybrids. A three-way hybrid (A × B) × C involves a single-cross and an inbred line in its pedigree. The single cross is used as female parent and the inbred line as pollen parent. Three-way crosses are used in cases where only three inbred lines which combine well are available. Recently, a hybrid having the pedigree (AU 151 × AU 102) × AU 71 has been released officially under the name of 'Shahbaz'.

iii. Double-cross hybrids. A double cross (A × B) × (C × D) is the hybrid progeny of a cross between two single crosses. Since double-cross seed is produced on highly productive single-cross plants, double-cross hybrid seed can be produced more economically than single-cross seed, which is produced on low-yielding homozygous parents. Double-cross hybrids are widely used in many countries. An example of a double-cross hybrid which is produced from local lines is AUH 817. This double-cross hybrid combines the characters of high yield and early maturity (90 days) and has been approved for general cultivation recently under the name of 'Quaid'. It has the following parents: (AU 239 × AU 82) × (AU 71 × AU 121).

iv. Double top-cross hybrids. A double top-cross (A × B) × V hybrid is the progeny from a cross between a single-cross and an open-pollinated variety. The single cross is used as a female parent and the open-pollinated variety as a male. Double top-cross hybrids are somewhat more variable than double crosses. One of the outstanding double top-cross maize hybrids developed locally is AUH 492, which is a cross between (AU 195 × AU 491) × 'Ehsan'. This is a white, short-duration hybrid which is high yielding and has good resistance to foliar diseases and stalk rot.

v. Multiple crosses. Sometimes the F_1 of six-way or eight-way multiple crosses may be used commercially. Usually multiple crosses are intended to produce synthetic varieties rather than using F_1 progeny. Advanced generations of these hybrids may be used after several cycles of random mating.

4.4.4 Improvement of existing lines

Sometimes a plant breeder wants to improve an elite inbred line which may be superior in many respects but deficient in one or two traits. The aim is to eliminate the weaknesses while maintaining all other qualities intact. This may be accomplished through **convergent improvement**. This method involves reciprocal improvement of two lines in a single cross. The F_1 is repeatedly crossed to each of the parental lines, so that the favourable genes from each of the parental lines are transmitted to each other in the succeeding generations through repeated backcrossing. Plants with desired characters and plant vigour are selected and used in developing each cycle of backcrossing. This procedure is carried out for about three or more generations, and selection is practised during backcrossing to retain as many of the genes contributed by the non-recurrent parent. By doing so, yield is improved in both recovered lines. In addition to backcrossing and convergent improvement, other methods of inbred line improvement include multiple convergence, sister-line development and gamete selection.

QUESTIONS

1. Discuss the weaknesses of mass selection. How would you overcome them in a breeding programme?
2. How would you differentiate a synthetic variety from varieties developed by mass selection?
3. Why is full-sib family selection considered twice as effective as half-sib family selection?
4. (a) What is recurrent selection? (b) How does one kind of recurrent selection differ from another? Explain your answer in terms of the gene action involved in each type.
5. In contrast to self-fertilized crops, a variety of cross-pollinated crops can not be developed from a single best plant. Explain why this is so.
6. Explain the genetic basis of inbreeding depression.
7. Discuss the advantages and drawbacks of a synthetic maize variety versus an F_1 hybrid for use in a developing country like Pakistan.
8. (a) Define combining ability. (b) Differentiate between general and specific combining ability.
9. (a) Define heterosis or hybrid vigour. (b) Explain the two hypotheses that best explain heterosis.

10. Define the following terms: top cross, polycross, progeny test, selection, full-sib, and convergent improvement.

REFERENCES

Bahadur, K. Progress from modified ear-to-row selection in two populations of maize (*Zea mays* L.). Lincoln, Nebraska: University of Nebraska; 1974. Ph.D. thesis.

Burton, G. W,; Gates, R. N. Using recurrent restricted phenotypic selection to improve bahiagrass and pearl millet forage yield (personal communication); 1991.

CIMMYT. PARC/CIMMYT (Center Internacional de Mejoramiento de Maiz y Trigo, International Maize and Wheat Improvement Centre) Collaborative Programme. Maize research and development in Pakistan. Mexico, D.F.: CIMMYT; 1989.

Comstock, R. E.; Robinson, H. F.; Harvey, P. H. A breeding procedure designed to make maximum use of both general and specific combining ability. Agron. J. 41:360–367; 1949.

East, E. M. Inbreeding in corn. Annual Report of the Connecticut Agri. Exp. Station for 1907–1908; 1908.

East, E. M. Heterosis. Genet. 21:375–397; 1936.

Gardner, C. O. An evaluation of effects of mass selection and seed irradiation with thermal neutrons on yield of corn. Crop. Sci. 1:241–245; 1961.

Gardner, C. O. Estimates of genetic parameters in cross-fertilizing plants and their implications in plant breeding. In Hanson, W. D.; Robinson H. F. eds. Statistical genetics and plant breeding: A symposium and workshop. Washington, D.C.: National Academy of Sciences, National Research Council; 1963:p. 225–252.

Gelaw, B. ed. To feed ourselves: A proceedings of the First Eastern, Central and Southern Africa Regional Maize Workshop, Lusaka, Zambia, March 10–17, 1985. 6600 Mexico, D.F. Mexico: CIMMYT; 1986.

Hallauer, A. R.; Miranda–Filho, J. B. Quantitative genetics in maize breeding. Ames, Iowa: Iowa State University Press; 1981.

Hallauer, A. R.; Sears, J. H. Changes in quantitative traits associated with inbreeding in a synthetic variety of maize. Crop. Sci. 13(3):327–330; 1973.

Hull. F. H. Recurrent selection and specific combining ability in corn. J. Am. Soc. Agron. 37:134–145; 1945.

Jenkins, M. T. The segregation of genes affecting yield of grain in maize. Agron. J. 32:55–63; 1940.

Jenkins, M. T.; Robert, A. L.; Findley, W. R., Jr. Recurrent selection as a method for concentrating genes for resistance to *Helminthosporium turcicum* leaf blight in corn. Agron. J. 46:89–94; 1954.

Jones, D. F. The effects of inbreeding and cross breeding upon development. Conn. Agric. Expt. Sta. Bull. 207:5–100; 1918.

Jones, D. F. Continued inbreeding in maize. Genet. 24:462-473; 1939.

Lonnquist, J. H. Modification of the ear-to-row procedure for the improvement of maize populations. Crop. Sci. 4:227–228; 1964.

Lonnquist, J. H.; McGill, D. P. Performance of corn synthetics in advanced generations of synthesis and after 2 cycles of recurrent selection. Agron. J. 47:319–323; 1955.

Moll, R. H.; Robinson, H. F. Quantitative genetics investigation of yield of maize. Zuchter 37:192–199; 1967.

Moll, R. H.; Stuber, C. W. Comparisons of response to alternative selection procedures initiated with two populations of maize (*Zea mays* L.). Crop Sci. 11:706–711; 1971.

Paliwal, R. L.; Sprague, E. W. Improving adaptation and yield dependability in maize in the developing world. Lisboa 27, Apdo Postal 6–641, 6600, Mexico, D.F.: CIMMYT; 1981.

Poehlman, J. M. Breeding field crops. 3rd ed. New York: AVI, Van Nostrand–Reinhold Co., Inc.; 1987.

Poehlman, J. M.; Borthakur, D. Breeding Asian field crops: With special reference to crops of India. rev. ed. Calcutta, India: Oxford and IBH Publishing Co.; 1969.

Shull, G. H. A pure line method of corn breeding. Am. Breed. Assoc. Rep. 5:51–59; 1909.

Singh, J. Current breeding concepts in the improvement of cross-pollinated crops. In Holmes, J. C.; Tahir, W. M. eds. More food from better technology. Proceedings of the 3rd FAO/SIDA Seminar on Field Food Crops in Africa and the Near East. Rome: FAO; 1983:p. 163–180.

Sprague, G. F.; Eberhart, S. A. Corn breeding. In Sprague, G. F. ed. Corn and corn improvement. Monograph 18. Madison, Wisconsin: American Society of Agronomy; 1977: p. 305–362.

Sprague, G. F.; Russel, W. A.; Penny, L. H. Recurrent selection for specific combining ability and type of gene action in yield heterosis in corn. Agron. J. 51:392–394; 1959.

Webel, O. D.; Lonnquist, J. H. An evaluation of modified ear-to-row selection in a population of corn (*Zea mays* L.). Crop Sci. 7:651–655; 1967.

Outline of Chapter 5

5. IMPROVEMENT OF ASEXUALLY PROPAGATED CROPS

Muhammad Siraj Swati
Hidayat-ur-Rahman[1]

LEARNING OBJECTIVES

Studying this chapter should enable a student to:

- Understand the characteristics of a clone.
- Discuss the advantages and disadvantages of asexual propagation of crops.
- Explain the value of mutation breeding for clonal crops.
- Explain why clonal crops often degenerate with time.
- Plan a breeding programme for asexually propagated crops.
- Understand and explain the role of tissue culture in propagation and improvement of asexual crops.

5.1 Introduction

Asexual propagation is that type of reproduction in which the population develops through mitotic cell division without fertilization, and the fusion of gametes (fertilization) does not occur. The plant population consists of identical individuals resembling the maternal plants and is called a clone. There are two types of asexual propagation: (1) from vegetative parts, and (2) from seeds.

Propagation from vegetative parts. In this type of propagation, parts such as bulbs, tubers, corms, rhizomes, bulbils, suckers, or setts are used as propagules (Chapter 2). These parts are described briefly here.

tuber – a swollen underground stem with buds or eyes from which shoots and roots can arise to develop into new plants, e.g. potato.

---

[1]Dr. Muhammad Siraj Swati and Dr. Hidayat-ur-Rahman are in the Department of Plant Breeding and Genetics, NWFP Agricultural University, Peshawar.

corm – a thick, solid, more or less rounded, underground stem with a few internodes. It is covered with a number of loose, sheathing scale leaves. In the axils of the scale leaves, one or more buds can, under favourable conditions, develop into aerial shoots. *Colocasia* and crocus, for example, reproduce by corms.

rhizome – a modified stem growing horizontally below the soil surface. It has nodes, internodes, buds, and leaves. When rhizomes are separated from the mother plant, the buds grow into aerial shoots under proper environmental conditions. Many grasses, ginger, and turmeric are common examples of plants that propagate by rhizomes.

bulb – a specialized underground shoot, in which the stem or axis is reduced to a short, broadly conical disc which gives off adventitious roots from its lower surface. On its upper surface, the disc bears an apical bud and some lateral or axillary buds. In the spring the axillary buds develop into aerial shoots to form daughter bulbs. Onions, garlic, and lilies are examples of bulbs.

sucker – an offshoot developed from a standing plant at its lower underground nodes, as in banana and date palm.

sett – a piece cut from the aerial stem which has buds at the nodal regions, as in sugarcane.

stolon – a stem which grows horizontally to the soil surface and forms roots and shoots at the nodal regions, as found in strawberry plants.

Propagation from seeds. In some species of higher plants, seedlike structures may develop by an asexual process called **apomixis** (Chapter 2). There are various forms of apomixis, but from the genetic point of view these can be divided into two types:

1. **OBLIGATE APOMIXIS:** In this case apomixis occurs regularly, completely replacing the normal sexual process of reproduction. Only clonal populations, which represent the genotype of the maternal plants, are generated.
2. **FACULTATIVE APOMIXIS:** In this type of situation, apomixis is *not* the sole method of reproduction, and is supplemented by sexual reproduction. Because the plants produce a mixture of both apomictic seeds and sexually-produced seeds, which exhibit genetic variation, facultative apomixis creates variation in the population through the process of recombination and segregation.

If the two forms of apomixis can be kept under control as with buffel grass (*Cenchrus ciliaris*), then the advantages of both the sexual and asexual mechanisms are obtained together. The sexual process is utilized to create variation, and the clonal combination is fully exploited by obligate apomixis.

5.2 Characteristics of clones

A **clone** is characterized by certain distinguishing features which differentiate it from plants developed through sexual reproduction.

1. All the individuals of a clonal population are genetically as well as phenotypically identical. The reason for this identity is their origin and development from a single plant through mitotic cell division which produces daughter cells that are genetically identical to the parental ones. The members of a clone, therefore, represent the parental genotype and also produce genetically uniform progeny. Plants propagated by vegetative means or through apomixis are clones. As long as there is no mutation, variation among clones is only environmental. Genetic variability in a clone arises only due to somatic mutations in buds, which happens very infrequently.

2. All the members of a clone are heterozygous. Although the member plants of a clonal population are phenotypically alike, genetically they are heterozygous. The parental plant in a vegetatively propagated crop is almost always heterozygous. This can be easily observed from the variation arising among the plants raised from the seeds of a clone. That is why improved varieties of fruit trees (e.g. mango), potato, and citrus are not raised from seeds but from vegetative parts. If clones are sexually propagated and successive inbreeding is followed, the plants will exhibit severe loss of vigour due to homozygosity.

3. Clones are stable. Clones are as stable as pure lines, and no segregation or variations occur in them or in their future generations. Even after many years of cultivation and vegetative propagation, a clone retains its original characters as such.

4. Because clones have a common ancestor and identical genotypes, the phenotypic variation within a clone is only environmental.

5. The phenotype of a clone is due to the effects of genotype and environment: the genotype (G) × environment (E) interaction, (G × E), over the population mean (μ). Thus the phenotypic differences among clones are partly due to their genotype and partly due to environment, i.e. the G and E components.

6. Theoretically, clones are immortal. A clone can be maintained indefinitely through asexual reproduction, but because they are genetically stable, clones usually degenerate due to viral or bacterial infections, and can become extinct because of their susceptibility to virulent new races of disease-causing organisms or adverse environmental conditions.

Characteristics of asexually propagated crops. Most asexually propagated crops share some common features.

1. A majority of them are perennials, e.g. sugarcane and fruit trees. The annual crops are mostly tuber crops, e.g. potato, cassava, and sweet potato.
2. Many of the asexually propagated crops are interspecific hybrids, e.g. banana and sugarcane.
3. Most of them show reduced flowering, and seed set may also be extremely low. They are invariably cross-pollinated. Some varieties do not flower at all.
4. They are characterized by a high degree of heterozygosity and show a tremendous amount of inbreeding depression.

5.2.1 Advantages of asexual propagation

1. It preserves the genotype of the parent indefinitely, irrespective of its genetic makeup.
2. Seedlings developed through vegetative means are generally more vigorous than those coming from true seeds because they have greater reserve food. In addition, being heterozygous, they often show heterosis.
3. Clonal populations are extremely uniform, with no variations except those which arise through mutation.
4. Plants propagated through vegetative means can be selected at any stage without the danger of changing their genetic composition in subsequent generations.
5. Upon sexual reproduction, their progeny give rise to a wide range of segregation variation.
6. Breeding of clonally propagated plants has the advantage of easily fixing the genetic variability in the progeny. The job of a breeder is to identify and isolate the desired genotype from a diversified, sexually propagated population, and then practice vegetative propagation.
7. The ability of a parent to transmit its characteristics in a cross (general and specific combining ability) can be fully exploited within the permissible range of selection efficiency.
8. If apomixis is controlled it can have the advantages both of vegetative propagation—fixation of genetic makeup (obligate apomixis)—and sexual reproduction—recombination of genes (facultative apomixis).
9. Some plants not normally reproduced through asexual reproduction can be quickly multiplied by using vegetative propagation (e.g. rice).
10. If the first-appearing shoots fail, new sprouts can grow from the tubers or other propagating parts, and loss of the entire crop is avoided.

11. Micropropagation by tissue culture is an efficient approach. It offers numerous opportunities for the vegetative production and multiplication of otherwise difficult species.

5.2.2 Disadvantages of asexual propagation

1. Cultivars developed through vegetative propagation degenerate quickly because of the accumulation of deleterious somatic mutations. Clonal populations are genetically uniform and degeneration through diseases is very fast. These populations are also very susceptible to environmental shocks. The relatively large size of the propagules (tubers, stem cuttings) used in vegetatively propagated crops makes them vulnerable to the buildup of soil-borne pathogens and virus infections.
2. If the propagules are stored for a long time and if storage conditions are not proper, they lose vegetative vigour and viability.
3. A larger volume of the product is required for propagation than when seeds are used. With potato, for example, 2.5 tonnes of tubers would be needed to seed one hectare, as compared to a few hundred grams of true seed. In developing countries, 20 percent of the potato crop is saved for seed, and thus unavailable for use as food. That is why the International Potato Research Centre is actively engaged in developing potato varieties from true potato seed (TPS). One such cultivar was released in 1982.
4. Clonal populations usually have a problem of lowered fertility, and to obtain all desired crosses is usually difficult.
5. Clonally propagated crops are often polyploids, aneuploids which are mosaics for chromosomal composition which lack regular diploid inheritance and manifest erratic segregation if bred sexually.

5.3 Sources of genetic variation within clones

In contrast to sexually propagated crops where segregation and recombination during meiosis result in increased genetic variability, asexually propagated crops lack this source of variation. However, there are some other means which lead to the creation of genetic variability in these crops. These include rare somatic mutations, mechanical mixture as human error, and occasional sexual reproduction.

Mutation. The frequency of somatic mutations (also called bud mutation) is extremely low, around 10^{-5} to 10^{-7}. They do, however, occur and thus create a detectable amount of genetic variability in asexually propagated crops. Usually dominant mutations are observed more frequently than recessive mutations. Bud mutations may lead to the formation of chimeras, individuals having segments which are different from the rest of the plant.

Somatic mutation makes possible the selection of buds to establish new clones. This approach has been very effective in the improvement of clonally propagated perennial crops like fruit trees and crops where flowering or seed formation does not occur. In order to detect desirable buds, a large population needs to be observed, as somatic mutations and chimeras arise by rare random mutation and not through a pre-planned hybridization programme.

Mechanical mixture. During the process of harvesting and handling, a certain amount of mechanical mixing with other genotypes by human or mechanical error can occur. This mixing can lead to genetic variation due to cross-pollination at flowering, and thus new genetic variation is introduced into the original clone.

Sexual reproduction. Occasional sexual reproduction does occur and leads to recombination and segregation. The seedlings obtained from sexual reproduction are genotypically different from the asexual progeny. Selection of superior genotypes is more effective in old clones, as with the passage of time and the accumulation of spontaneous mutations they become more variable than new clones.

5.4 Clonal degeneration

Theoretically, clones can be maintained indefinitely without any genetic change. But in practice, clones have a limited life and often older clones lose vigour and productivity. This loss in vigour and productivity with time is known as clonal degeneration. It is assumed that clonal degeneration is an inherent phenomenon; however, detailed studies show that it results from several factors, including exposure to environmental stresses and accumulation of deleterious genetic variations.

Mutation. Small genetic changes or mutations occur as a recurring process. They can become a problem, because many of them can be harmful and their accumulation over a long period of time produces detrimental effects on the plant's development and growth. Since the clones have no mechanism to eliminate the deleterious mutations, they continue to accumulate in the population. Furthermore, some mutations occur at a relatively high frequency. For example, in some potato clones certain mutations occur at a frequency of 10^{-3} (which is quite a high range), and are a serious problem in the maintenance of these clones.

Viral diseases. Viruses are perhaps the major cause of clonal degeneration. Viral diseases are easily transmitted through vegetative propagules and hence they spread with time in clonal crops. Many potato clones have been lost because of viral infections.

Bacterial diseases. Bacterial infections are also a cause of clonal degeneration. For example, the ratoon stunting disease of sugarcane is caused by a bacterium *Clavibacter*$_{xyli}$. There are other similar instances where a bacterium is responsible for the degeneration of clones.

5.5 Methods of improvement of vegetatively propagated plants

Various methods are practised for the improvement of vegetatively propagated crops.

5.5.1 Introduction or acclimatization

In this important method of plant improvement, a number of lines are imported from abroad, generally from areas with similar environmental conditions. The germ plasm thus made available is then tested for climatic adaptation; its performance is compared to that of the standard cultivars of the area. This method was very successful with sugarcane in the NWFP, where during the period from 1960 to 1986 several cultivars of sugarcane ('CP 44–101', 'CP 48–103', 'CP 65–357', 'L62–96', etc.) were released for commercial cultivation. In Punjab, the process of acclimatization was also successful in introducing the 'Triton' sugarcane cultivar.

5.5.2 Selection

Selection as a method of plant improvement is probably older than written history. In this method, several hundred to several thousand clones are procured and screened to find the ones suited to local agro-climatic conditions. Any clone which shows superiority in overall productivity or any single agronomic characteristic may be directly released as a variety or noted for use in future breeding programmes. The direct selections are multiplied and tested in replicated trials. If the plant proves its worth, it is then released as a commercial variety.

During the earlier stages of clonal selection for breeding programmes, selection is based on characters with high heritability using single plants or single plots containing an unimproved clonal population of plants. Most of the plants raised through asexual means are comparatively weak when reproduced by seed through sexual mechanisms. One obvious reason is the poor storage potential of the propagating material itself, but in addition to this the homozygosity of deleterious recessive alleles also plays its role. Emphasis is thus placed on elimination of weak and undesirable plants or clones. Since at these early stages the full expression of all traits may not be possible to judge, identification and selection for overall superior genotypes can be unreliable. To give it a fair chance, the material is grown for further generations and eventually selected on the basis of replicated trials. Utilizing a large population of seedlings is advisable to enable proper selection of desired types. However, during the selection process the population size is rigorously reduced for subsequent generations. At this stage, emphasis is on identifying and selecting superior clones.

Proper selection for characters with low heritability and high environmental influence can then be done by controlling for environmental variables. Finally, comparisons in yield and other quality characters are made between the new variety and standard commercial cultivars. If improvement is obtained for certain important characters without clear-cut inferiority for others, then the new variety can be released commercially.

Vegetatively propagated plants are easily affected by pathogens and have a comparatively short commercial life. Therefore, methods resulting in quicker multiplication of propagules should be adopted. Tissue culture techniques can be used for faster multiplication and for the production of healthy seed. Further trials can be conducted to pinpoint geographical areas where healthy seed can be routinely produced for normal cultivation. In the case of potato plants, 'seed potatoes' are produced at higher altitudes where infection is minimum. The seed potatoes are then supplied for crop cultivation in the plains.

The various steps involved in clonal selection are briefly described below.

1. First year. From a mixed population, a few hundred to a few thousand desirable propagules are selected. The mixed population can be a land race, a local variety, or a segregating population. Since wild or unimproved populations will usually be mixtures of heterozygous clones, it may be possible to isolate clones superior for particular characteristics which are more influenced by the environment. Rigorous selection can be practised for highly heritable traits. Weak and less vigorous clones are eliminated. Selection for characters with low heritability can be delayed until later generations.

2. Second year. Selected plants are planted vegetatively in spaced rows. Subsequently, if the propagation material becomes available in sufficient quantity, a replicated yield trial is conducted, otherwise the material is grown for further multiplication and during this period a more rigorous selection can be done. The selection is primarily based on phenotypic observations, and clones showing apparent weakness are eliminated. Consequently, the number of clones is drastically reduced. Based on visual scoring and superior performance for the desired characteristics, 50–100 individuals are advanced to the next generation for further evaluation.

3. Third year. A replicated preliminary yield trial is conducted along with a suitable commercial variety as a check for comparison. Clones with superior performance and desirable characteristics are selected for multi-location trials.

4. Fourth to eighth years. Replicated yield trials are conducted at several locations along with suitable checks. Rigorous selection is done for yield potential, quality, and disease resistance. Clones with superior qualities are identified.

5. Ninth year. The superior clones are multiplied and released as varieties.

5.5.3 Sexual methods

Sexual methods are adopted for the improvement of vegetatively propagated crops when a source of variation is desired. As vegetatively propagated plants can be selected at any stage, the selected clones are generally heterozygous in nature. This is due to the existence of positive correlation of vigour with heterozygosity. It is therefore expected that even the F_1 plants may give segregation for some gene pairs.

a. Selfing or inbreeding. Self-pollination or inbreeding is the fertilization of the ovary by the male flower of the same plant. This procedure is utilized to get rid of deleterious recessive alleles which are expressed only in homozygous conditions. After crossing, the inbred lines obtained by selfing may result in healthy heterozygous combinations, which can be selected if found desirable.

b. Crossing of standard commercial varieties. When crossed, commercial varieties can form segregating populations. The seed developed from such plants can be planted. The seedlings are then passed through different screening procedures and the elite forms can be recommended for commercial cultivation.

c. Poly-cross method or melting-pot procedure. In this case a large number of cultivars are brought together in an isolated area and natural cross-pollination is permitted to occur. Rearrangement of the pots or planting arrangement will enhance randomness and thus the diversity of cross-pollination. The resulting seed can be traced only to its maternal parent, as the male parent is not known due to the natural cross-pollination among the selected clones in isolation. The seed thus obtained can then be used in a breeding programme until the desired level of homozygosity is achieved.

d. Modified backcrossing. In this procedure a desired character from the donor parent is obtained through the backcrossing procedure. The conventional backcrossing procedure is, however, modified by using a different standard line for each backcross. The purpose of this backcrossing is to concentrate desirable genes to the extent that it does not adversely affect plant vigour, while retaining heterozygosity. Any individual clone which appears desirable can then be picked out for routine screening procedures.

e. Interspecific crossing. Vegetatively propagated plants such as sugarcane have undergone so much intercrossing that the varieties under commercial cultivation are all hybrids resulting from different species. Most often *Saccharum spontaneum* and *S. barberi* are used as a source of disease resistance and cold tolerance. In this procedure, the improved variety which is deficient in the desired character, i.e. disease resistance or cold tolerance, is crossed with a genotype of *Saccharum spontaneum* or *S. barberi* which has the desired alleles. Then the genes of *S. officinarum* are concentrated by repeated backcrossing with noble canes, called **noblization**. In this way the genetic variability of a crop is increased, and can later be used as a source or superior clones if selection is done.

5.5.4 Mutation breeding

Mutation breeding is of special importance in the improvement of vegetatively propagated plants, especially when whole propagules are exposed to mutagenic sources such as X-rays, gamma rays, and neutrons, or even to chemical mutagenic agents which are applied to actively growing buds. It is also useful in obligate apomictics. It is well known that the adaptability, yield, and quality characters in crop plants show complex genetic inheritance. Difficulties are thus encountered in crossing vegetatively propagated crops, which makes mutation breeding an ideal choice when the goal is to improve some simple characteristic(s) without introducing gross changes.

Konzak (1984) describes some 297 mutants of potato and sugarcane which have been developed for commercial use. With sugarcane, the mutants were selected to improve various characters such as the suppression of flowering, early maturity, higher yield, and resistance to insect pests and diseases. Potato mutants were developed which are insensitive to photoperiod and early maturity, and thus are able to grow in regions to which other forms are not adapted.

5.5.5 Tissue culture

Plant tissue culture has been successfully applied in various ways to improve agricultural crops. It is actually a highly refined technique of vegetative propagation of plants, in which different plant parts can be used to regenerate whole new plants. Single cells or small pieces of plant tissue are aseptically made to regenerate into new plants in suitable culture media and environmental conditions. Meristem culture has been utilized in the production of disease-free clones of many important vegetatively propagated plants such as potato, cassava, sweet potato, and yam. The method has thus enabled some clonally propagated commercial varieties to regain their value after approaching extinction due to the development of pathogenic viruses. Tissue-culture techniques are now used as a standard procedure for the supply of disease-free seed in some countries. Strawberry is propagated from meristem tips to get virus-free stocks.

Tissue culture is also used as a means of quick multiplication for making available a bulk supply of commercial forms to growers as well as commercial propagation of hardy horticultural plants like orchids. In addition, this technique is useful in preserving germ plasm to avoid the extinction of precious genetic material. Preservation of tissues, especially meristems, is valued where seed production is not easy, seed is viable only for a limited period, is highly heterozygous and not a true representative of the cultivar, or is unsuitable because of pathogens (Kartha 1984).

More recently, the possibility of fusing isolated protoplasts (cells without the cell wall) of completely unrelated species has opened up an entirely novel approach to crop improvement. Tissue culture, especially callus culture, is applied in the production of mutants or in screening for stress. Polyploidy and aneuploidy are the most common phenomena occurring during *in vitro* culture. Sugarcane breeders have successfully utilized this method in the production of cultivars resistant to eyespot disease, downy mildew, and sugarcane mosaic caused by viruses (Cailloux 1984). Increase over the parental variety in cane yield, tillering capacity, and sucrose content were also noted (Kartha 1984).

5.6 Problems in breeding asexually propagated crops

Although the breeding of asexually propagated crops seems simple, in actual practice the procedure is very complicated. There are three major problems in breeding clonal crops: (1) difficulties in genetic analysis, (2) perennial life cycle, and (3) reduced flowering and fertility. In any hybridization programme, estimates of general and specific combining ability are very helpful in selecting suitable parents to be crossed. Genetic analysis in most clonal crops is very difficult because of their low seed set or complete sterility, limited flowering, and perennial life cycle, which tremendously increases the time span required to obtain sexual progeny to use in the evaluation of clones.

5.6.1 Reduced flowering and fertility

Almost all of the clonal crops grown for vegetative parts have a tendency to reduced flowering, and some varieties of sugarcane, potato, and sweet potato do not flower at all. Clonal crops grown from vegetative plant parts often face problems in flowering behaviour. Flowering is either completely suppressed or they have problems in fertility because most of them carry imbalanced chromosomes and functional gametes are not formed. These crops also show a considerable amount of sterility due to their polyploid nature and complex genomic makeup, as well as cytoplasmic factors. Polyploidy and aneuploidy also present complex problems of segregation of genes. As a result, it may not be possible to arrange crosses within desired parents, not only because flowering is erratic but also because the flowering times do not synchronize. The result is that many desirable clones cannot be used in hybridization. The reduced flowering and high sterility seem to be the result of a response which is correlated with long-term selection for increased yield of vegetative parts.

5.6.2 Problems in sugarcane breeding

Sugarcane is an important crop of Pakistan, the breeding of which suffers from the above-mentioned problems. Tropical in origin, the plant does not flower away from its homeland. Moreover, in commercial planting flowering is not desirable because it leads to substantial reduction in sucrose content. Selection is therefore usually practised against the flowering character.

Supply of fuzz. Prior to Independence, the Sugar Cane Research Station at Coimbatore, India, supplied **fuzz** (the seeds and floral structures of sugarcane) to breeders in various locations. After Independence, when the fuzz supply was stopped, researchers in Pakistan adopted other means of meeting their requirements. Earlier efforts made at higher altitudes at Charrapani near the Murree Hills were successful in the production of viable fuzz, but the site did not provide an environment conducive to flowering in all varieties of sugarcane. At Faisalabad successful production of viable fuzz was obtained with daily irrigation and increase of day length by giving extra artificial light using incandescent bulbs. But in spite of initial progress in generating fuzz locally, most improvement work has been based on the introduction and acclimatization of exotic cultivars. This has succeeded to a considerable extent at the Sugar Crops Research Institute, Mardan in NWFP. However, Sindh and Punjab suffered a lot and tried to generate their own fuzz locally. The results were quite encouraging; however, as an alternative they also based their improvement work on imported fuzz supplied by different international sources such as Barbados, Louisiana (USA), Bangladesh, and Brazil. At Faisalabad, these attempts succeeded in developing some varieties ('BL4', 'BL19', 'LU116', 'L118', and 'BF/162') which were also adaptable in Sindh.

However, on the whole the programme was not very successful and suffered from limitations. (1) The supply of seed was not regular. (2) The crossing programme of the source country was not oriented to the special requirements of our country. (3) The quantity of fuzz sent was not sufficient to fulfil the demand for effective selection. It is estimated that there is only a 1 in 100,000 chance of obtaining a sugarcane seedling having desirable characters that can lead to development of a cultivar. Breeding is a game of chance, and the probability of superior combination is reduced when the total population on which selection is based is small. With potato, according to one estimate, only one successful cultivar will result from a population of about 1,000,000 seedlings.

Vigorous efforts are being made to handle the problem of viable seed production. The Pakistan Agricultural Research Council is concentrating its efforts in Dargai, Malakand Agency, where profuse flowering is observed under natural conditions. Viable true seed is being produced from the native material at the site, and a regular crossing programme is under way for the development of new varieties of sugarcane according to the requirements of the region. The Sugarcane Research Institute at Faisalabad has also planned to establish a station for seed purposes at the coastal areas of lower Sindh where the environment is suitable for flowering. With such facilities established, the country will be less dependent on imported germ plasm.

Induction of flowering. Flowering can be induced in sugarcane by controlling the photoperiod and temperature.

PHOTOPERIOD: The number of hours of daylight and their distribution are the most important factor influencing flowering in sugarcane. At the equator, where the day length remains about 12 hours year round, sugarcane flowers all year round. There seems to be a strong relationship with the diurnal changes as well as with the temperature. In fact, the ratio of day to night length is more important than the absolute number of hours of light. Adjustments in diurnal variation can thus induce flowering in most varieties. Interruption of the dark or light period even for a very short time delays flowering. This fact is sometimes exploited in order to synchronize flowering.

TEMPERATURE: A temperature of 21–25°C is required for pollen viability and seed setting. A drop to below 18°C adversely affects pollen viability. The female portion of sugarcane does not show similar sensitivity to low temperatures and can be used from open fields.

In addition to photoperiod and temperature, poor soils, high humidity, and high elevation encourage flowering. Some chemicals like CMU (3–para–chbrophyml–1, 1–dimethylurea) are effective in suppression of flowering.

QUESTIONS

1. (a) What is asexual propagation? (b) Why are populations developed through this procedure genetically uniform?

2. (a) What are the two main types of asexual propagation? (b) Why do some of the seed formed develop into clonal populations? Explain.

3. Describe the different types of propagule used in asexual propagation. Give an example of a plant which reproduces with each of these forms.

4. (a) What is apomixis? (b) Discuss some of its advantages over vegetative propagation. (c) How is it possible to combine the advantages of sexual and vegetative propagation in a single method?

5. Describe the characteristics of asexually propagated plants. Why do some plants reproduce by this method?

6. What are the special advantages associated with asexual propagation? Describe them in detail.

7. Discuss some of the disadvantages of asexual propagation? How can some of them be overcome through the adoption of special techniques?

8. What features of a clone differentiate it from a population developed through sexual means?

9. (a) What is meant by clonal degeneration? (b) What factors cause a clone to degenerate and result in the running out of a commercial variety?

10. What methods are used for the genetic improvement of vegetatively propagated plants? Describe the different steps in the development of a variety.

11. (a) How can sexual methods be employed in the improvement of vegetatively propagated plants? (b) What difficulties are involved and what potential solutions are available?

12. Why are wider crosses more successful in breeding vegetatively propagated plants than with sexually reproducing ones?

13. Flowering is generally erratic in plants propagated through vegetative means. How can this difficulty be overcome with sugarcane plants?

14. (a) What factors help in the induction of flowering in sugarcane? (b) Indicate certain areas in Pakistan where the natural environment is conducive to flowering in sugarcane.

15. Flowering in sugarcane is affected both by genetic makeup and environment. Discuss this statement.

16. (a) How can viable seed be produced in sugarcane? (b) Describe some measures to prolong the viability of sugarcane seed.

17. What is the special role of recent technological developments in induced mutations and tissue culture in breeding asexually reproducing plants? Illustrate your statements with specific examples.
18. Tissue culture is not only a method of propagation but also a technique of raising healthy seed and creation of variability. Discuss this statement.
19. Write a detailed note on the different uses of tissue culture. Discuss the different variations of this technique and their special uses.
20. What are the various problems involved in breeding of vegetatively propagated plants?

PRACTICAL EXERCISES

1. Suggest a scheme which can help in distinguishing true seeds from apomictic ones. What difficulty will there be if in a certain species apomixis is facultative rather than obligate?
2. Collect different types of propagules used for vegetative propagation and try to determine what should be done to use these economically and effectively for the development of clonal populations.
3. Record different characters such as plant height in populations raised through vegetative and sexual means, noting the variability in each of the two cases. Where will you expect more variation and why?
4. Visit the different areas of your country where sugarcane blooms or potato sets true seed. Compare the environmental conditions of these localities with those of your own area.
5. Visit a tissue culture laboratory to understand the different steps involved—starting from the selection and multiplication of tissues and resulting finally in the regeneration of plants. Do you find these plants more attractive than those raised through ordinary means?
6. If you can find a sugarcane plant in bloom, try to investigate the viability of its pollen by applying carmine stain.
7. Test the viability of sugarcane fuzz using some of the available techniques in your lab.
8. Arrange a visit to a lab engaged in the study of viral diseases and observe the symptoms of viral disease in a potato tuber. Make this observation both with the naked eye and with an electron microscope.

REFERENCES

Agnihotri, V. P. Diseases of sugarcane. New Delhi: Oxford and IBH Publishing; 1983.

Alexander, A. G. Sugarcane physiology. New York: Elsevier Scientific Publishing Co.; 1973.

Allard, R. W. Principles of plant breeding. New York: John Wiley and Sons; 1960.

Ayub Agricultural Research Institute, Faisalabad. Annual Report 1983–84. Lahore: Directorate of Agricultural Information, Deptt. of Agriculture, Govt. of Punjab; 1985.

Brandes, E. W.; Sartoris, G. B. Sugarcane: its origin and improvement. USDA Yearbook of Agriculture; 1936.

Cailloux, M. Plant tissue culture: rapid propagation, induced mutations and the potential role of protoplast techniques. In: Vose, P. B.; Blixt, S. G. eds. Crop breeding: A contemporary basis. Oxford: Pergamon Press; 1984:p. 311–346.

Chaudhari, H. K. Elementary principles of plant breeding. 2nd ed. New Delhi: Oxford and IBH Publishing; 1986.

Clements, H. F. Sugarcane crop logging and crop control: Principles and practices. London: Pitman Publishing; 1980.

Cocking, E. C.; Riley, R. Application of tissue culture and somatic hybridization to plant improvement. In: Frey, K. J. ed. Plant breeding II. Ames, Iowa: The Iowa State University Press; 1981:p. 85–116.

Fehr, W. R.; Fehr, E. L.; Jessen, H. J. eds. Principles of cultivar development, Vol. 1. New York: Macmillan; 1987.

Hanna, W. W.; Bashaw, E. C. Apomixis: its identification and use in plant breeding. Crop Sci. 27(6):1136–1139; 1987.

Hoopes, R. W.; Plaisted, R. L. Potato. In: Fehr et al. eds. Principles of cultivar development. Vol. 2. Crop species. New York: Macmillan; 1987:p. 385–436.

Humbert, R. P. The growing of sugarcane. New York: Elsevier Publishing Co.; 1968.

Kartha, K. K. Tissue culture techniques for virus elimination and germplasm preservation. In: Janick, J. ed. Plant breeding reviews, Vol. 2. Westport, Connecticut: Avi Publishing; 1984:p. 265–282.

Khan, M. I. Prospects of sugarcane breeding in the N.W.F.P. Province. In: Khan Rana, M. S.; Sarwar. M. eds. Proceedings of workshop on agricultural characters and morphological description of sugarcane varieties, held at NARC, Islamabad. Islamabad: NARC, Pakistan Agric. Res. Council; 1989:p. 38–42.

Khokhlov, S. S. ed. Apomixis and breeding (translated from Russian). New Delhi: Amerind Publishing; 1970.

Konzak, C. F. Role of induced mutations. In: Vose, P. B.; Blixt, S. G. eds. Crop breeding: a contemporary basis. Oxford: Pergamon Press; 1984:p. 216–292.

Malik, K. B. Constraints in cane breeding under Pakistan conditions. In: Khan Rana, M. S.; Sarwar. M. eds. Proceedings of workshop on agricultural characters and morphological description of sugarcane varieties, held at NARC, Islamabad. Islamabad: NARC, Pakistan Agric. Res. Council; 1989:p. 18–37.

Mayo, O. The theory of plant breeding. Oxford: Clarendon Press/Oxford University Press; 1980.

Mendoza, H. A.; Sawyer, R. L. The breeding programme at the international potato centre (CIP). In: Russell, G. E. ed. Progress in plant breeding, Vol. 1. London: Butterworths; 1985:p. 117–137.

Mohammad, I. Heritability of flowering in sugarcane. Louisiana State University; 1986. Ph.D. dissertation.

Mohammad, I.; Rehman, G.; Qayyum, A. Two new promising varieties of sugarcane for Peshawar. Frontier J. Agri. Res. 8:108–113;1983.

Nayyar, M. M. Sugarcane cultivation in Pakistan. In: Khan Rana, M. S.; Sarwar. M. eds. Proceedings of workshop on agricultural characters and morphological description of sugarcane varieties, held at NARC, Islamabad. Islamabad: NARC, PARC; 1989:p. 1–17.

Nickell, L. G. Plant growth regulators in the sugarcane industry. In: Mc Laren, J. S. ed. Chemical manipulation of crop growth and development. Proceedings of the Easter School in Agriculture Science, University of Nottingham. London: Butterworth Scientific; 1982:p. 167–189.

Poehlman, J. M. Breeding field crops. 3rd ed. Westport, Connecticut: Avi Publishing Co.; 1987.

Poehlman, J. M.; Borthakur, D. Breeding Asian field crops. New Delhi: Oxford and IBH Publishing Co.; 1969.

Rhoades, R. E. The incredible potato. Nat. Geog. 161: 668–694;1982.

Rowe, P. Breeding bananas and plantains. In: Janick, J. ed. Plant breeding reviews, Vol. 2. Westport, Connecticut: Avi Publishing Co.;1984:p. 135–155.

Russell, G. E. ed. Progress in plant breeding. Vol. 1. London: Butterworths; 1985.

Simmonds, N. W. Principles of crop improvement. London: Longman; 1979.

Singh, B. D. Plant breeding. New Delhi: Kalyani Publishers; 1983.

Stevenson, G. C. Genetics and breeding of sugarcane. London: Longmans, Green; 1965.

Welsh, J. R. Fundamentals of plant genetics and breeding. New York: John Wiley and Sons; 1981.

Outline of Chapter 6

6. NEW ADVANCES IN PLANT BREEDING

Khushnood A. Siddiqui[1]

LEARNING OBJECTIVES

After working with this chapter, a student should be able to:

- Discuss basic concepts of the theory and practice of mutation breeding.
- Utilize mutation breeding as a tool for the improvement of different crop species.
- Analyze plant genetic material at the biochemical and molecular levels.
- Utilize *in vitro* techniques in contemporary and future plant breeding.
- Develop a synergetic approach for the construction of an ideotype.

6.1 Introduction

Scientific plant breeding started with Gregor Mendel (1822–1884). The years that Mendel worked with his pea plants gave birth to genetics and caused a revolution in science. New advances in plant breeding gained momentum with the help of continuous inputs from biophysics, biochemistry, microbiology, and bioengineering.

Plant breeding is an evolutionary process. In fact, our present day crop plants are the products of evolution. During the last 90 years technological innovations in genetics and related disciplines have made it possible to direct this evolution by utilizing hybridization, heterosis, polyploidy, mutations, wide crosses, tissue culture, haploidy, protoplast fusion, somaclonal variation, antibody probes, and DNA probes. Comprehensive crop improvement can only be accomplished through a synergetic approach involving all the conventional and non-conventional methods of plant breeding (Siddiqui et al. 1991, Karpinski et al. 1994). This chapter on new advances will be mainly devoted to modern, non-conventional methods of plant breeding.

---

[1]Dr. Khushnood A. Siddiqui is Chief Scientific Officer and Director, Atomic Energy Agricultural Research Centre, Tandojam, Sindh.

6.2 Mutation breeding

Mutation is a sudden hereditary change that is not due to a recombination of genes. In a strict sense, a mutation implies the chemical alteration of a gene or a small structural chromosome change. Changes in genes are often called **point mutations** (Elliott 1958). Changes in chromosome number usually involve either increase or decrease in the number of chromosome sets. Mutations also take place due to addition or subtraction of less than a whole set of chromosomes. Inversions, translocations, deletions, and duplications are additional sources of chromosomal mutations. In a broader sense, mutations can be classified into two main categories: spontaneous mutations and induced mutations (Siddiqui 1990). Being the ultimate source of all genetic variability, mutations have played a vital role in organic evolution. The use of mutations in plant breeding is popularly known as **mutation breeding** (Siddiqui et al. 1991).

Mutation research today is completely dominated by investigations of experimentally induced mutations. During the last 30 years mutation breeding has gained pronounced momentum. This is a global activity, which is regularly monitored by the Joint FAO/IAEA Division of Nuclear Techniques in Food and Agriculture, International Atomic Energy Agency, Vienna, Austria, mainly through symposia, seminars, research contracts, and publications. Future possibilities of mutation breeding have also been discussed in detail, particularly in the context of new emerging biotechnologies (Siddiqui 1990, Siddiqui et al. 1991).

The objectives of mutation breeding must be clear and well-defined. The success of mutation breeding will depend on the availability of genetic information about the starting material. The theoretical expectation of inducing mutations in crops will depend on, among other factors, the relative proportion of genes with positive and negative effects and the degree to which the genes of the parental genomes operate as a balanced set. Induced mutations increase in frequency with decreasing magnitude of change, and the probability of improvement also increases with decreasing magnitude of change (Gregory 1965). Induced mutations are particularly useful when it is desired to improve one or two easily identifiable characters in an otherwise well–adapted variety (Siddiqui et al. 1989).

6.2.1 Tools of mutation breeding

1. Physical mutagens. Induction of mutations is often achieved through physical mutagens such as X-rays, gamma rays, fast neutrons, thermal neutrons, ultraviolet radiation, and beta radiation (Elliott 1958). All of these types of radiation (with the exception of ultraviolet rays) ionize atoms in a tissue by detaching electrons from the atoms (IAEA 1977). Ionization from irradiation with X-rays, gamma rays, and beta rays occurs sparsely along the track of the ionizing particle. The ion density is expressed in **linear energy transfer (LET)**, which is the energy dissipated per unit length along the tracks of the ionizing particles (IAEA 1977).

EFFECT OF OXYGEN AND WATER CONTENT. The presence of oxygen is important in changing and enhancing the effect of irradiation in biological systems (Singh et al. 1986). In tissues with low water content such as dry seeds, the most commonly irradiated plant part, the radiation-induced radicals decay slowly. If the water content of the seeds is high, the radicals decay rapidly. So in irradiating seeds, the breeder must ascertain the water content and oxygen status in order to obtain repeatable results. Within plant cells, certain regions are more sensitive to radiation than others. These regions, called radiosensitive targets, correspond generally with the location of DNA in the cells. Irradiation can affect either one or both of the DNA strands of the double helix. If the breaks involve a single strand, the linear integrity of the DNA molecule is still intact and repair of the break restores normal DNA. When the breaks involve both of the strands, they are not as readily repaired; this is the most important radiation induced damage of DNA resulting in detectable mutations.

2. Chemical mutagens. In addition to physical mutagens, a large number of chemical mutagens are also used to induce mutations in crop plants. Although the number of new chemical mutagens employed is continuously increasing, only a few are really useful in practical mutation breeding. Most of these belong to the special class of alkylating agents and include ethyl methanesulphonate (EMS), diethylsulphate (DES), ethyleneimine (EI), ethyl nitroso urethane (ENU), ethyl nitroso urea (ENH), and methyl nitroso urea (MNH).

A. ALKYLATING AGENTS. The alkylating agents react with DNA by alkylating the phosphate groups as well as the purine and pyrimidine bases. Another group of useful mutagens are base analogues. They are closely related to the DNA bases, and can be incorporated into DNA in place of normal bases without hindering its replication. Because they differ in certain ways from the normal bases, these mutagens can give rise to pairing errors in DNA which result in mutations. An example of a frequently used analogue is 5 bromouracil (BU). A related compound, maleic hydrazide (MH), causes chromosome aberrations.

B. AZIDES. Azides were known to be effective only in diploid wheat, barley, and peas. Earlier efforts to use sodium azides were unsuccessful in polyploid wheats with 28 or 42 somatic chromosome numbers (Sigurbjornsson 1983). However, recently azides such as sodium azide have been effectively and successfully used in inducing mutations for qualitative (Siddiqui et al. 1989) as well as for quantitative traits (Siddiqui et al. 1991) in hexaploid wheat ($2n = 6x = 42$).

3. Treatment of plants or plant parts. Whole plants or any part of the plant can be treated by radiation (Sigurbjornsson 1983). However, because of the physical bulk, special installations are needed to irradiate whole plants or large plant parts. Gamma fields are expensive to build and maintain. For treating vegetative parts, such as branches of fruit trees or entire plants, smaller gamma chambers with controlled environmental conditions are useful. Gamma irradiation facilities are available in Pakistan at Tandojam, Faisalabad, and Peshawar.

Seeds are most commonly used for irradiation. They offer a number of advantages. They are easy to handle and store, and can be maintained for extended periods of time in a vacuum almost free of oxygen as well as under high pressure of oxygen or other gases. They can be subjected to a variety of physical and chemical environments; they can be desiccated, soaked, heated, or frozen. When dry, seeds are almost inert biologically and severe environments cause no significant biological damage.

A great number of studies have been done on the different biological, environmental, and chemical factors which can modify the effect of radiation treatment. These factors modify mutagenic effectiveness (the number of mutations per unit of dose) and mutagenic efficiency (ratio of the number of mutations to damage such as sterility, stunted growth, etc.) of both physical and chemical mutagens in cells of higher plants. Detailed information on different aspects of mutagenesis is available in *Manual on Mutation Breeding* (IAEA 1977).

6.2.2 A generalized scheme for mutation breeding

A generalized scheme for mutation breeding is illustrated in Figure 6.1. The successive generations derived from mutagen-treated seeds or plants are designated as M_1, M_2, M_3, M_4, and so on. The M_1 generation starts with the germination of mutagen-treated seeds. In this generation only dominant mutations will be expressed. The M_1 plants will be heterozygous for newly induced recessive mutant genes. Newly induced recessive mutations will be expressed and observed in the M_2 generation following self-fertilization of the M_1 plants. Identification and isolation of mutants in cross-fertilizing species require controlled pollination when producing the M_2 generation. In self-incompatible species, homozygous mutants can only be expressed in the M_3 generation following mating of sister M_2 plants. The M_2 generation provides the best opportunity of selection, which can be for yield components, disease resistance, improved quality, heat tolerance, drought tolerance, and improved plant characteristics such as shattering resistance (Siddiqui et al. 1981) and stem stiffness (Siddiqui 1983).

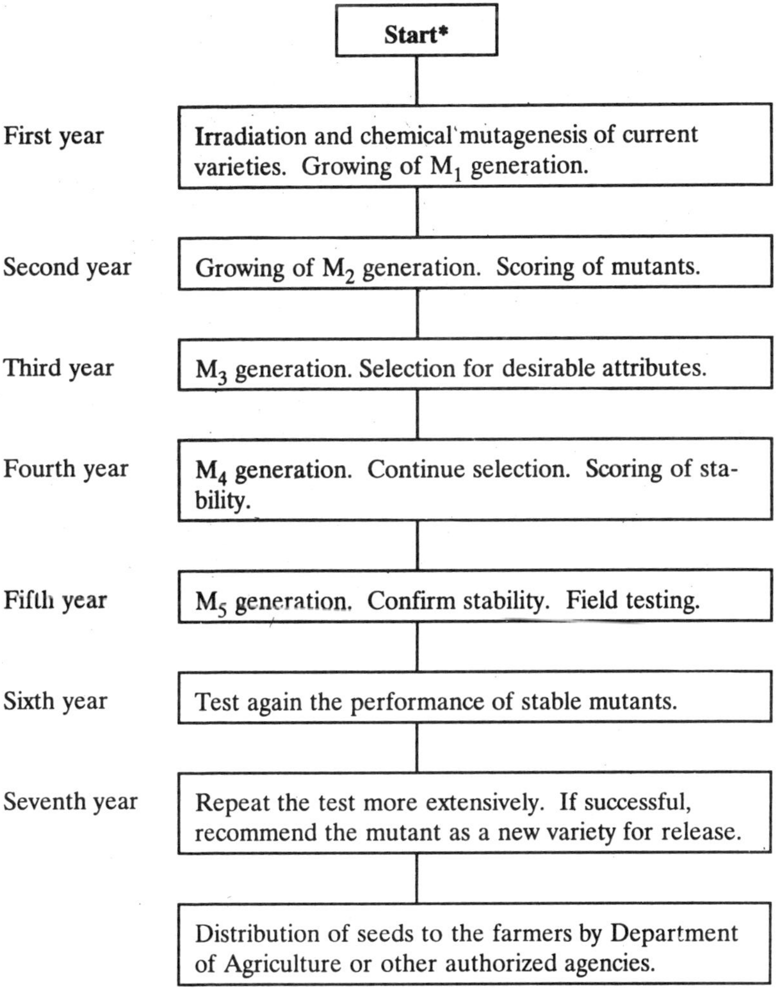

* This is a continuous process.

Figure 6.1 A generalized scheme of mutation breeding in crop plants

The effectiveness of selection will depend upon the efficiency of screening techniques. Gross morphological mutations such as changes in plant height (Fig. 6.2, 6.3) or glume colour (Fig. 6.4) can be detected visually in the field or greenhouse. In fibre crops like cotton, visual selection is quite reliable (Shuaib et al. 1981) if morphological mutants exhibit differential boll formation (Fig. 6.5) in the M_2 generation.

Morphological mutants find phenotypic expression at various stages of ontogeny depending upon the original attribute of the mother cultivar. For instance, in root crops such as sweet potato, the mutant character is discernable only after harvest (Fig. 6.6). Likewise, in cereals (Siddiqui and Arain 1986, Siddiqui et al. 1987) improved grain characteristics can only be detected after post-harvest processing or cooking (Fig. 6.7).

Figure 6.2 Rice mutant EF-I with reduced height (right) isolated from the treated population (gamma rays 250 Gy) of Basmati 370 (left).

Physiological characters such as early or late flowering (Fig. 6.8) can be easily identified. In fruit crops such as apples, however, one has to wait for years to find marked differences in the skin colour of the fruits (Fig. 6.9). For such characters as disease resistance, pest resistance, drought tolerance, and salt tolerance, other techniques are required for isolating real mutants.

Biochemical methods are needed for detection of attributes of mutants such as protein content and lysine content (Siddiqui 1972c, Siddiqui and Doll 1973, Siddiqui et al. 1975, 1978a). Mutants can be isolated in the lab through zymographic methods alone or in combination with Osborn fractionation, amino acid analysis, gel chromatography, and electrophoresis (Siddiqui et al. 1972). In recent years, availability of sophisticated equipment (Fig. 6.10) has greatly facilitated the isolation and characterization of mutants in the laboratory. Screening for characters such as salt tolerance can be facilitated by the use of certain 'marker' characters like proline accumulation (Siddiqui 1983). Likewise, deviants with new functions such as mutations which give resistance to the phytotoxicity of a systemic fungicide can be accurately detected with appropriate techniques (Siddiqui and Haahr 1971, Siddiqui 1972d).

Confirmation of mutation is usually accomplished in the M_3 generation. In the M_4 generation, stable mutants are evaluated in field trials for polygenic characters such as yielding ability. When planning experiments, it

Figure 6.3 Mutagenesis in soybean. Heavy bearing determinate mutant (left) developed via fast neutron treatment (15 Gy) of mother variety 'Improved Pelican' (right).

should always be kept in mind that positive mutations are rare. Large populations are, therefore, needed to increase the probability of finding

Figure 6.4 Successful azide mutagenesis in hexaploid wheat (2n = 6x = 42) (0.2M, 6 hr soaking dose + 3 hr treatment). Mother cultivar has brown glumes, while Mutant 233 has golden glumes.

positive-yield mutants. There are mainly two different selection methods for yield mutants (Siddiqui 1983). The first method is concerned with the selection and isolation of mutants for qualitative characters. The yielding ability of selected mutants is then tested in subsequent generations.

The second method is direct, in which the progeny testing starts in the M_3 generation. If the material is stable and non-segregating, station trials are conducted in the M_4 generation. In the M_5 generation, the attributes of the mutants are further evaluated in the so-called station trials. In Tandojam, we have successfully confirmed high yielding mutants by growing the M_5 generation in the balanced lattice design (Fig. 6.11). The same statistical technique is used in M_6 for re-confirmation of mutants endowed with polygenic positive attributes. Mutants with above average performance are promoted in the zonal trials. These trials are conducted in agriculturally important regions in which a particular crop is well adapted. The purpose of zonal trials is, therefore, to test the adaptability of a particular mutant across different environments. Some mutants may be specifically adapted to a particular agricultural environment; whereas others will be widely adaptable (Siddiqui et al. 1979). Zonal trials are conducted for two years. After this, successful mutants are promoted in the national trials, which are also conducted for two years at around 50 locations throughout the country.

Figure 6.5 Peaceful use of atomic energy. 'Cluster mutant' bearing cotton bolls in cluster of three (right) derived from the treatment with gamma rays (200 Gy) from Cobalt^{60} at AEARC, Tandojam.

During these trials a mutant remains under constant vigilance for stability, uniformity, distinctness, and its reaction to different diseases and insect pests. Simultaneously, mutants are also evaluated for their water and fertilizer requirements. In addition, agronomic trials are also conducted to determine their suitability to different soil types and planting time.

The purpose of this exercise is to evolve a new cultivar. The proposal for a candidate cultivar/variety which is submitted to the Provincial Seed Council contains detailed experimental data on quantitative as well as qualitative characteristics of economic importance.

6.2.3 Successful products of mutation breeding

Mutation breeding is now an established and practical tool of crop improvement. The information available in 1991 from the joint FAO/IAEA Division, Vienna and from other sources shows that more than 1300 varieties (Fig. 6.12) have been developed and released through mutation breeding.

Another contribution of mutation breeding which is often overlooked and underestimated is the production of useful germ plasm. New realities of industrialization and other environmental hazards are resulting in significant genetic erosion, depletion, and extinction of natural genetic resources.

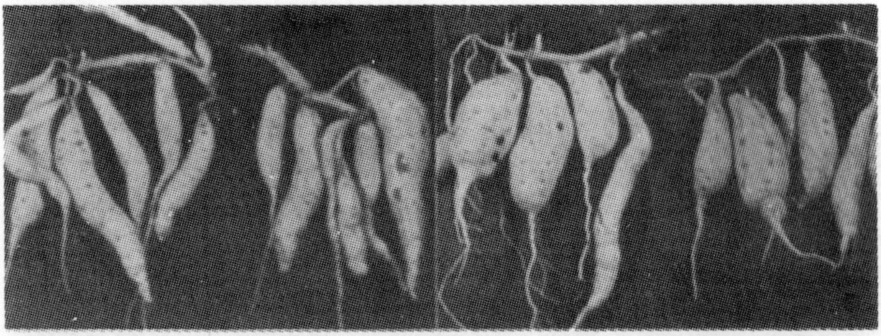

Figure 6.6 Skin colour mutation in sweet potato (*Ipomoea batata* Lam.). Photograph courtesy of Dr. Kukımura, H. Instt. of Radiation Breeding, Ibaraki, Japan.

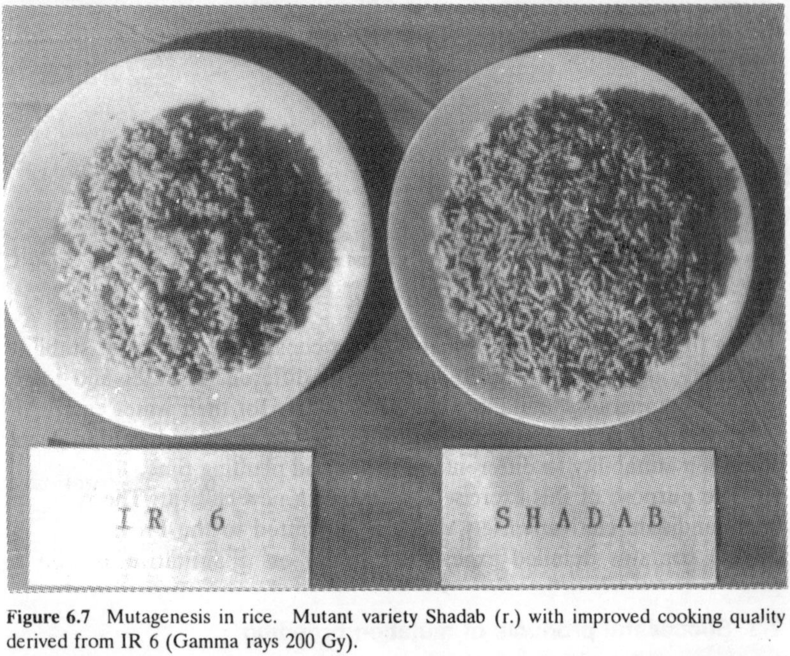

Figure 6.7 Mutagenesis in rice. Mutant variety Shadab (r.) with improved cooking quality derived from IR 6 (Gamma rays 200 Gy).

Contemporary mutation breeding is playing a significant role in creating new germ plasm containing genes endowed with new dimensions and even with new functions (Siddiqui and Haahr 1971, Siddiqui 1972d, Siddiqui 1990). Mutants which are not released as cultivars are often utilized in fundamental genetic, biochemical, and physiological studies and also in recombination breeding (Siddiqui et al. 1988).

Illustrations from crops of economic importance. One of the most successful examples of utilization of induced mutations for plant breeding

Figure 6.8 Induction of early maturity following mutagenesis in rice variety IR 6.

a b c

Figure 6.9 Induction of fruit colour mutation.

is provided by wheat (Siddiqui 1972b, 1983, 1990; Siddiqui et al. 1978b, 1979). Before 1960, there is no evidence of release of any variety through the use of induced mutation (Table 6.1). One of the main reasons for this is the fact that earlier geneticists like Stadler (1928a, b) were convinced that induced mutagenesis would not result in improved plant cultivars. This pessimism was shared by a geneticist who said, 'The probability of improving a naturally adapted genotype by mutation breeding is equal to the probability of improving a mechanically sound typewriter by dropping it to the ground from the top of a four-storey building.' (Quoted in Frey 1965:467).

Figure 6.10 SE 600 Dual Cool vertical slab unit. Example of biochemical method used for isolation of mutants in the laboratory.

This was mainly due to lack of understanding of the genetic constitution of bread wheat (2n = 6x = 42 = AABBDD) (Siddiqui 1972a) and its subsequent manipulation through mutagenesis (MacKey 1960, Siddiqui and Jafri 1986), selection, hybridization, and biotechnological innovations (Siddiqui 1990).

Successful synergetic approaches have resulted in a large number of mutant wheat varieties from different parts of the world, and at present mutation breeding in wheat is a significant global phenomenon (Table 6.1). These new cultivars represent a wide variety of improved traits of economic importance (Fig. 6.13).

Use of induced mutations has been very successful in rice breeding (Fig. 6.14) particularly in Asia (Table 6.2). Mutation breeding in rice has also enormous future possibilities (Siddiqui et al. 1991).

Mutation breeding in cotton (Table 6.3) has been economically very profitable in Pakistan (Saeed Iqbal et al. 1991). The cotton variety 'NIAB–78' has given an additional return of more than 800 million Pakistani rupees in a single year, thus repaying three times the total of 18 years of investment in research (Fig. 6.15) (Siddiqui et al. 1991).

Use of induced mutations in legumes not only provides new cultivars in crops like chickpea and mung bean (Table 6.4) with improved yield and nutritive quality, but also genotypes which have more efficient symbiosis and nitrogen fixation and are suitable for multiple cropping under sustainable agricultural systems.

6.2.4 New advances in mutation breeding

Mutation breeding has enormous future possibilities (Siddiqui et al. 1991). One of the newer methods through which these possibilities can be realized is popularly known as the *in vitro* technique of mutation breeding. According to information published by FAO/IAEA, *in vitro* techniques can be applied in various steps of the breeding process, and offer a number of advantages for mutation breeding (Siddiqui 1990). The following are the main applications of *in vitro* techniques in mutation breeding.

Mitotically active cells and meristematic tissues can be propagated *in vitro* for the preparation of suitable material for mutagenesis. The meristematic dome will give rise to mericlinal chimerae after mutagenesis. Periclinal chimerae and/or homohistont shoots can be obtained by repeated *in vitro* subculture of axillary buds. Diplontic selection can be controlled by modifying the *in vitro* conditions; for example composition of media. When genetic variation is induced by mutagenesis before or during culture, haploid individuals provide an opportunity for direct selection of mutations; *in vitro* culture of plant cells and tissues can be instrumental in obtaining variants with alterations in genomes, chromosomes, and genes, even without any mutagenic treatment.

6.3 New advances in plant cytogenetics, and their relevance to modern plant breeding

During the last 30 years, significant advances in plant cytogenetics have provided new inputs to contemporary plant breeding, resulting in the genetic improvement of diploids as well as polyploids (Tsuchiya 1986). In diploids, chromosome mapping has provided radical plant breeding techniques such as the BTT (balanced tertiary trisomics) method developed by Ramage (1964). The duplication method was developed by Hagberg (1962) and further refined by Hagberg and Hagberg (1978). In maize and tomato, structural changes such as reciprocal translocations and deficiencies have been successfully used in cytogenetic linkage mapping.

Barley trisomics. Barley breeding has been greatly accelerated by the new cytogenetic advances. New approaches, especially trisomic analysis, have been successfully used in the chromosomal mapping of barley (Tsuchiya 1986).

Figure 6.11 Confirmation of induced mutations in the field. The author (right) with his associates.

Chromosome mapping in rice. New cytogenetic advances have provided additional information which can be exploited in breeding new varieties of rice. These advances include the assignment of genes and linkage groups to specific chromosomes in *japonica* rice and *indica* rice through the use of trisomics. Recent information indicates that the genus *Oryza* contains diploid (2n = 24) species with genomic constitution AA (*Oryza sativa*), BB (*Oryza punctata*), CC (*Oryza officianalis*), and EE (*Oryza australiensis*). The information on the diploid species having a DD genome is not available, but three tetraploids namely *Oryza latifolia* (CCDD), *Oryza alta* (CCDD) and *Oryza grandiglumis* (CCDD) are supposed to possess the D genome and are designated accordingly. More extensive cytogenetic investigations are needed to clarify the genomic relationships in the genus *Oryza*. There seems to be no dispute on the genomic designation of *Oryza minuta*, *Oryza malampuzhaensis*, and *Oryza schweinfurthina*. These three tetraploids all have the same (BBCC) genetic constitution. Among the diploid cereals, *Oryza sativa* (2n = 24) has the lowest amount of DNA (0.6 pg). It has been recently suggested that the prospects of conducting experiments involving 'chromosome walking' are brightest for rice (Siddiqui et al. 1991).

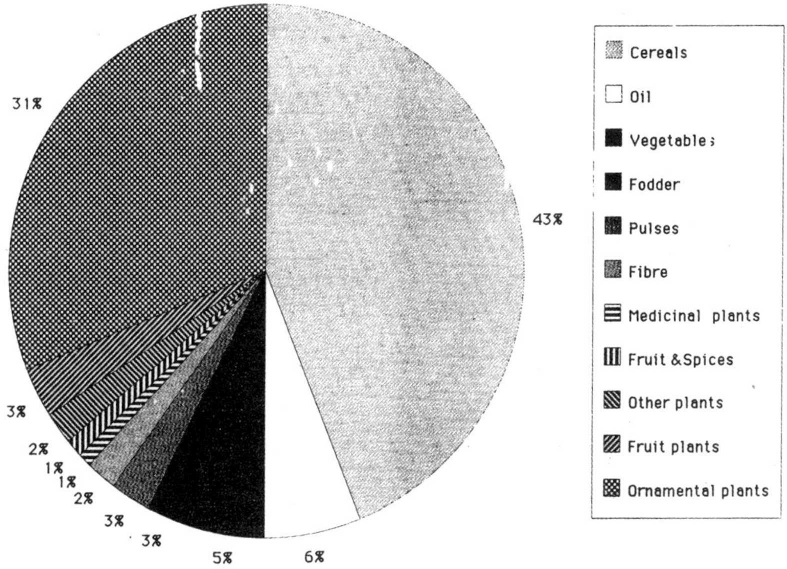

Figure 6.12 Percentages of 1300 varieties of crop plants developed through mutation breeding

Telomeric sequences in rye. In rye (*Secale cereale*), removal of telomeric heterochromatin sequences has been shown to be significantly related to decreased production of aberrant endospermic nuclei, which in turn resulted in an increase in 1000 kernel weight and test weight measured at maturity under field conditions (Gustafson and Kaltsikes 1986).

Diploids under stress conditions. The new synergetic approach (Siddiqui et al. 1991) focusses attention on genome monitoring. This is particularly important in diploid crops such as pearl millet which often grow under unfavourable conditions of infertile soils, intense heat, and limited rainfall. Such diploid crops cannot tolerate the additional stress of chromosomal aberration (Aslam and Siddiqui 1986) and therefore cannot maintain morphological uniformity and genetic stability across stress environments.

Meiotic behaviour of pigeon pea. Another diploid which has assumed considerable importance under stress conditions is pigeon pea (*Cajanus cajan*, 2n = 22). Cytogenetic methods have been used successfully to elucidate the genomic relationships between *Cajanus cajan* and seven different species, namely *Atylosia albicans*, *A. sericea*, *A. scarbaeoides*, *A. cajanifolia*, *A. lineata*, *A. grandifolia*, and *A. latisepala* of the related genus *Atylosia*. The new information suggests the presence of two nucleolar organizers per genome in both the genera. This information can be manipulated in the

The seed of the mother variety was treated with fast neutrons. This is a classical example of peaceful uses of atomic energy.

Figure 6.13 Jauhar 78, the first high-yielding, shattering-resistant and amber-grained wheat variety released in Pakistan.

future genetic improvement of pigeon pea.

Meiotic behaviour of chickpea. Regular meiosis appears to be the cytological characteristic of the genus *Cicer*, which includes the commercial chickpea, *Cicer arietinum* (2n = 16). Information is also available on the chiasmata frequencies of six other species of *Cicer*, namely *C. bijugum*, *C. chorassanicum*, *C. judaicum*, *C. pinnatifidum*, and *C. reticulatum*. All these species also have regular bivalent formation (2n = 16) and are readily amenable to utilization in chickpea improvement.

New polyploids. Since the discovery of the action of colchicine on the spindle of a dividing cell (Blakeslee and Avery 1937), many experimental polyploids have been produced. These include autopolyploids, formed by multiplication of sets of chromosomes within the limit of a species, and amphiploids, derived from the combinations of two distinct species. A good example is provided by the so-called man-made cereal, the hexaploid

Figure 6.14 Rice variety Shadab developed through mutation breeding released for general cultivation in Sindh.

Triticale (2n = 6x = 42 = AABBRR) derived by the combination of a tetraploid wheat species (e.g. *Triticum durum*, 2n = 4x = 28 = AABB) and a diploid species of cultivated rye, *Secale cereale* (2n = 2x = 14 = RR). Numerous variations of polyploid themes are now recognized.

During the 1970's, earlier methods of induction of polyploidy, viz. the crown method (Sears 1941), capping method (Bell 1950), and the tiller method (Cauderon and Saigne 1961) were modified by Siddiqui (1971a) for induction of polyploidy in multigeneric hybrids belonging to the tribe Triticeae. During the 80's, induced amphiploidy became an important method of crop improvement (Siddiqui 1982a). New amphiploids were produced in the genera *Avena*, *Hordeum*, *Linum*, *Lolium*, *Pennisetum*, *Festuca*, *Gossypium*, *Lotus*, *Vigna*, *Phaseolus*, and *Nicotiana*. More recently, polyploidy has been induced in *Solanum americanum* (Siddiqui and Jafri 1986) and in *Fragaria* × *ananassa* (Evans 1986).

In spite of continuous efforts during the last 90 years, genetic analysis of polyploids through the use of conventional genetic methods has not been successful (Siddiqui 1972a). The most significant development to the cytogenetics of polyploids has been made by Sears (1969), which is illustrated by production of monosomics (2n = 41) and nullisomics (2n = 40) in bread wheat (2n = 42). This development has made it possible to associate some visible morphological effects with the absence of a chromosome or a pair of

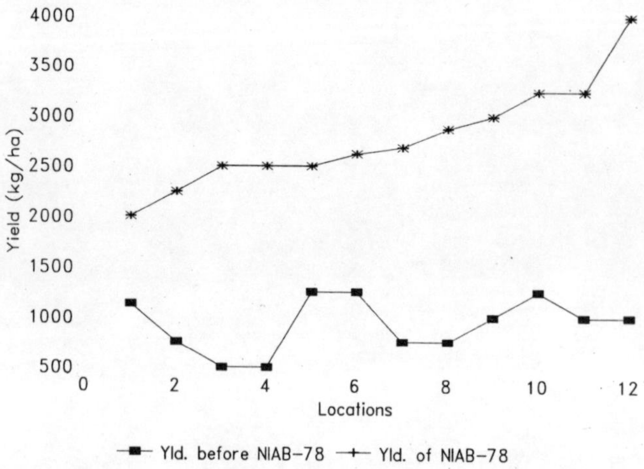

Fig. 6.15: Impact of NIAB-78 on the cotton yield of farmers' fields in Punjab(1986-89)

chromosomes. Monosomics and multisomics have permitted the use of chromosomal substitution as a new method in plant breeding.

Chromosomal location of genes. Monosomic, multisomic, and substitution methods have enabled plant geneticists to establish the location of genes on the chromosomes. The amount of information on gene locations has multiplied rapidly (Morris 1980). For instance, the catalogue of gene symbols for wheat alone has remained under constant revision during the last 20 years by the international coordinator for compiling the catalogue of gene symbols of wheat and related species (McIntosh 1988). Also, the compendium of alloplasmic lines and amphiploids has been completed separately (Maan and Gordon 1988). The fourth compendium of wheat–alien chromosome lines was published in 1988 (Shepherd and Islam 1988), and the fifth will be available for publication in 1993.

In addition to the published information, unpublished information on current developments includes the chromosomal location of the enzymes (esterase, glucophosphate-isomerase, α-amylase, and β-amylase) in wheat varieties and induced mutants, namely C591, CS, Jauhar 78, Yecora, H–68, M–192, Barani 70, Pak 70, Sind 81, Nayab, P–89, Sarsabz, Mexi–Pak, M–46, M–169, M–175, M–172, Pavon, 6134 × C 271 (Siddiqui et al. unpublished). In summary, substitution lines have been successfully used for identification of chromosomes concerned with protein content and quality (Siddiqui 1972c;

Siddiqui et al. 1978a). An elegant use of whole chromosome substitution lines has been in tracing the course of mutations in an allopolyploid (Siddiqui and Jafri 1986). However, the most significant contribution to the new plant breeding has been the utilization of the combined cytogenetic and statistical techniques developed by C.N. Law and his associates at Cambridge, England that make possible the detection and precise mapping of quantitative genes as well as those with qualitative effects (Law et al. 1983).

Figure 6.16 Phenotypic expression of Norin-10 based dwarfing genes in wheat.

New catalytic agents in plant breeding. The new cytogenetic developments have accelerated plant breeding activities throughout the world (Siddiqui 1982b). For instance, over the past two decades, global wheat production has risen at an annual rate of 3.4%. In most cases dramatic increases in wheat production became possible only with the concurrent widespread utilization of Norin 10 sources (Fig. 6.16) of dwarfing genes singly (Rht_1) and in appropriate combinations, e.g. $Rht_1 + Rht_2$ (Siddiqui 1982b, Gale and Youssefian 1985) (Fig. 6.17).

Such wheat varieties have stiff straw and are not prone to stem collapse or lodging before harvest. Over the past 15 years a considerable amount of information has become available on dwarfing genes in wheat (Gale and Youssefian 1985). The Rht_1 and Rht_2 have been shown to be members of a homologous series located about 14 recombination units from the centromeres of the α-arm of chromosome 4A and the short arm of 4D (McVittie et al. 1978).

Extraction of ancestral constituents of natural polyploids. Deletion of genomes is one of the most recent developments in plant cytogenetics (Shigenaga 1986). It has been studied as 'extraction' (Siddiqui 1963, 1964, 1969; Mochizuki and Shigenaga 1964), 'reconstitution' (Kerber 1964), and 'depolyploidization' (Carlbom 1969). Two different methods (Figs. 6.18 and

6.19) are employed. Implications and applications of these methods are discussed in detail by Siddiqui (1972a) and Shigenaga (1986).

Genetic bridges over infertility barriers. New advances in cytogenetics have also been instrumental in overcoming the genetic isolation of wild relatives of crop plants (Siddiqui 1982a). Most of these species contain useful genes of agricultural significance but do not cross readily with their cultivated relatives, thus making the incorporation of desirable genes difficult or impossible (Siddiqui 1977). A good example of this phenomenon is provided by *Aegilops ventricosa* (DDM^vM^v). This wild relative of bread wheat shares the D genome (responsible for breadmaking quality) with bread wheat (AABBDD). *Aegilops ventricosa* also contains genes for resistance to root rot and Hessian fly. Before 1964, no viable hybrid involving *Aegilops ventricosa* with bread wheat was known.

Figure 6.17 Dosage effects of height-reducing genes in wheat.

Genetic bridges were successfully employed by using amphiploids involving *Aegilops ventricosa* with genomic constitution $AADDM^vM^v$ and $AABBDDM^vM^v$ (Siddiqui 1964, Siddiqui and Jones 1967a). Similarly, amphiploids were used as bridges for transfer of alien genetic variation from *Aegilops sharonensis*, *A. longissima*, and *A. speltoides* (Siddiqui 1964, 1971a, 1977, 1982a). Likewise, problems of crossability were resolved by using autotetraploids of rye and monosomics of wheat (Siddiqui and Jones 1968b).

Synthesis of multigeneric hybrids without embryo culture. During the last 30 years synthetic amphiploids have been successfully used for the production of a large number of bigeneric and trigeneric hybrids without resorting to embryo rescue (Siddiqui 1964, 1971a-b, 1982a; Siddiqui and Jones 1967b, 1968a, 1969).

Chromosome segments from wild relatives. Chromosome segments from

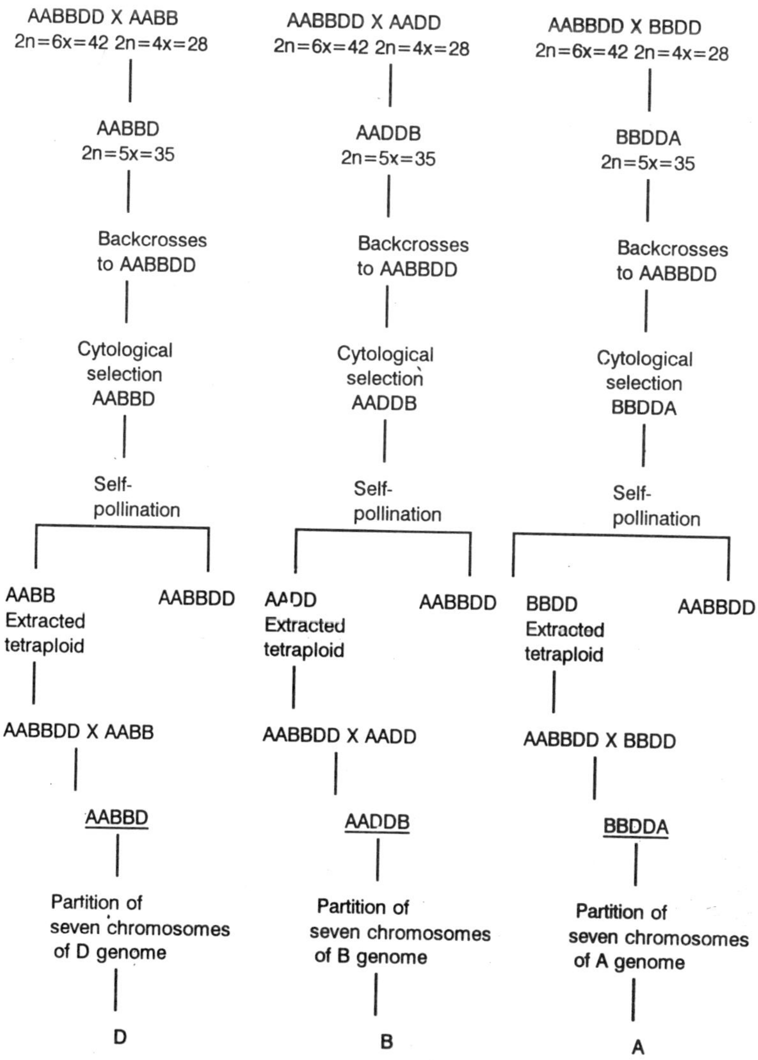

Figure 6.18 Schematic representation of the extraction method (Siddiqui 1963, 1964, 1969, 1972a).

wild species have been used in the transfer of characters of economic importance such as disease resistance in tomato (Rick 1982). Induced deletions, e.g. phlb (Sears 1977), allow homologous pairing and transfer of alien

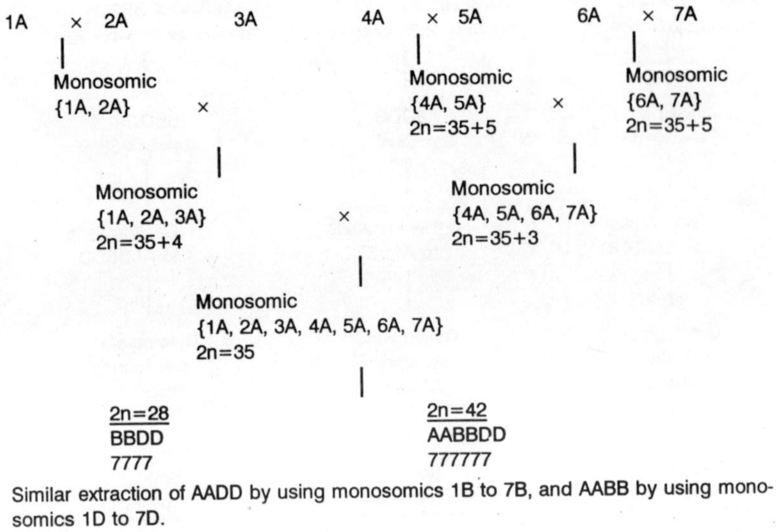

Figure 6.19 Proposed scheme for extracting tetraploid components of an allohexaploid by the monosomic method (Siddiqui 1963, 1964, 1972a).

chromosomes bearing useful genes. Another induced deletion, the Cornerstone male sterility mutants (mslc), has also been used in wheat breeding (Driscoll 1984).

Nucleo-cytoplasmic heterosis. Cytoplasmic relationships (Tsunewaki 1980) have been elegantly elucidated in recent years. Kihara (1986) has provided conclusive evidence that the heterotic effects of cytoplasm over the expression of nuclear genes can be exploited in future plant breeding.

6.4 *In vitro* techniques in contemporary plant breeding

In vitro technique is a blanket term used in the past to describe experiments in a cell-free system. In recent years, the meaning of the term has been modified to include the growth of cells under tissue-culture conditions. *In vitro* techniques are assuming increasing importance in contemporary plant breeding. The most important *in vitro* techniques which are being currently used in breeding are: tissue culture, cell-suspension culture, protoplast culture, anther culture, ovule culture, *in vitro* pollination–fertilization, embryo culture, micropropagation, and cryopreservation. The routes and aims of vegetative propagation are illustrated in Fig. 6.20.

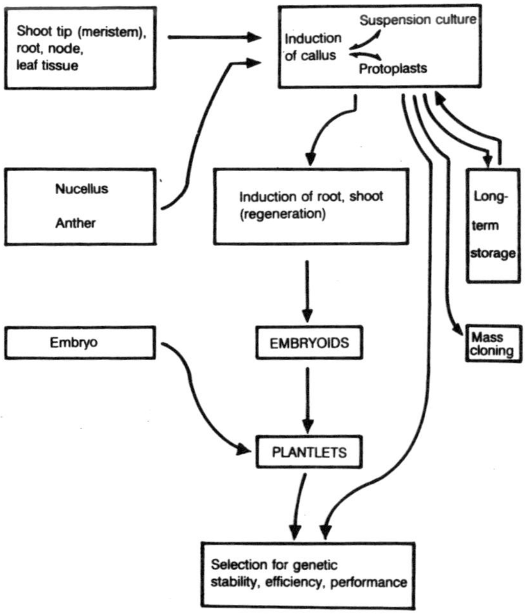

Figure 6.20 Schematic diagram of *in vitro* techniques currently employed in plant breeding (Courtesy: C. J. Jensen, Risö, Denmark).

6.4.1 Tissue culture

The term *plant tissue culture* is commonly used to describe the *in vitro* and aseptic cultivation of any plant part on a nutrient medium. The technique is simple and remains essentially unchanged since the time when it was first outlined by Haberlandt in 1902 (Krikorian and Berquam 1969).

Plant tissue culture technology is based on three fundamental principles. (a) The plant part (explant) must be isolated from the rest of the plant body. (b) The explant must be maintained in a controlled, preferably defined, nutrient medium. (c) Asepsis must be maintained. Cell culture and tissue culture have provided an efficient selection tool for modern plant breeding (Fig. 6.21).

1. Callus culture. With the culturing of an explant (an excised and isolated piece of tissue) on a nutrient medium, an unorganized, proliferating mass of cells appears which is known as a **callus**. With age, such a callus may show meristematic islands or strands, and individual or groups of tracheids and pigmented cells. This callus is transferred onto different media to regenerate plants. Regeneration of plants through callus culture provides genetic variability which is an important prerequisite for successful plant

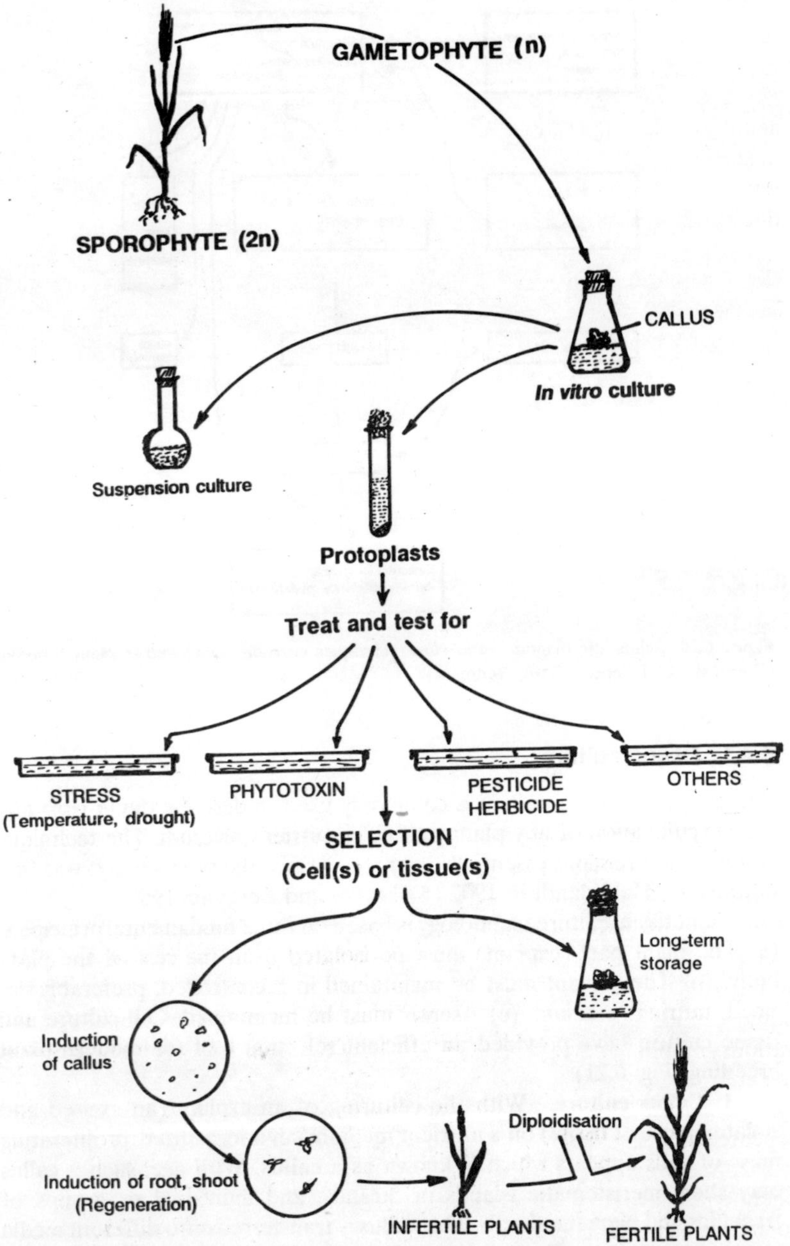

Figure 6.21 A new tool for modern plant breeding: selection via cell and tissue culture (Drawn by Habib-ul-Rahman Mian, following C. J. Jensen, Risö, Denmark).

breeding. It also saves time and labour by facilitating selection of novel genotypes.

2. Somaclonal variation. Callus culture is mutagenic in the broad sense, as callus cells maintained *in vitro* over a long period are usually cytologically unstable and after regeneration give rise to plantlets which are often characterized by genetic variability. This is popularly known as **somaclonal variation**, variation induced in plants regenerated through tissue culture.

Many agriculturally useful variants have been obtained through somaclonal variation (Table 6.5). In sugarcane a variant (70–6132) had enhanced cane yield (32%), increased sugar yield (34%), and higher stalk number (6%) than the donor F 164 (Liu and Chen 1978). Similarly, many clones showing resistance to Fiji disease, downy mildew, eye spot, and whip smut disease have been derived through callus culture (Krishnamurthi 1974, Larkin and Scowcroft 1981, Liu and Chen 1982). Many variants of maize have been obtained (Cowen et al. 1990). Somaclonally induced earliness and disease resistance have also been recorded in *Brassica juncea* (Katiyar and Chopra 1990).

6.4.2 Cell-suspension culture

Cell-suspension culture is the culture of isolated cells or very small cell aggregates dispersed in a liquid medium. The cell suspension is obtained by agitating pieces of callus in a liquid medium on a gyrating shaker. Cell-suspension culture is mostly used to study the physiological, biochemical, and metabolic characteristics of plant cells. It is also used to isolate cells showing tolerance to different stresses such as phytotoxins, salts, herbicides, metals, and drought conditions. Cell-suspension culture is particularly useful in inducing genetic variability in cells through different physical and chemical mutagens and also in the production of secondary metabolites (e.g. alkaloids like nicotine, purine, indole, etc.), proteins, and plant growth regulators.

1. Protocol for cell-suspension culture. Cell-suspension culture involves three main steps: (1) transfer of obviously friable sectors of callus to a liquid medium; (2) reduction of the size of the aggregate cells by applying various treatments, e.g. (a) the use of culture flasks with baffles, (b) forcing the culture through a sieve or by pumping the culture repeatedly through a syringe with a wide bore canula; (3) maintenance of cell suspensions by repeated sub-culturing.

2. Achievements of cell-suspension culture. Many agriculturally useful phenotypes have been selected through cell suspension culture for resistance to herbicides (Ranade and Feierabend 1991), pathotoxins (Katiyar and Chopra 1990), NaCl tolerance (Nabors et al. 1980), cold tolerance (Dix and Street 1975), and drought tolerance (Heyser and Nabors 1979).

6.4.3 Protoplast culture

One of the most significant developments in the field of plant tissue culture during recent years has been the isolation, culture, and fusion of protoplasts (Cocking 1972). The term **protoplast** refers to all the components of a plant cell excluding the cell wall. The techniques are especially important because of their far-reaching implications in studies of plant improvement, cell modifications, and somatic hybridization. Protoplast culture is used for the induction of mutations and genetic variability and also for the production of somatic hybrids. This process is very successful, particularly in those species where sexual hybridization is impossible or very difficult. Protoplast culture is also utilized for incorporation of viruses, bacteria, exogenous DNA, transplantation of nuclei, and implantation of chloroplasts.

1. Protocol for protoplast isolation, culture, and fusion. The successful isolation of protoplasts depends on many factors such as source tissue, plant species and variety, enzyme mixture, and physiological status of the plant. The protoplasts are usually isolated from cultured cells or leaf mesophyll cells by treating with enzyme solutions. The isolated protoplasts may be used to regenerate the plants directly, or for the production of somatic hybrids through fusion. Somatic hybridization can be achieved by using polyethylene glycol (PEG) or electroporation. PEG (Junker et al. 1987) as well as electroporation (Ou–Lee et al. 1986) can lead to both stable integration of introduced genes, or to expression of genes into the plant genomes (Lee et al. 1990). Protoplasts play an important role in plant transformations. Many approaches to plant transformation are currently being developed, but from the practical point of view relatively simple and reproducible methods involve *Agrobacterium tumefaciens,* and tumour-inducing (Ti) plasmids and their interaction with protoplast systems (Power and Chapman 1985). Protoplast technology is being constantly revised to obtain **cybrids** (cytoplasmic combinations). A recent success in protoplast technology is the stable transformation of maize after gene transfer by electroporation (Fromm et al. 1986).

2. Achievements through protoplast culture. Somatic hybrids have been produced between a large number of plant species, but most of these species combinations are restricted to the Solanaceae (Fladung 1990). Populations of plants developed through protoplast fusion have shown more variability than the variability available through sexual hybridization. Genotypic variability has been observed between different plants for traits such as plant height, leaf shape, leaf size, leaf petiole length, flower length, flower colour, pollen viability, crossability, and isoenzyme banding pattern (Evans 1983).

In conventional breeding, desired traits are selected through a recurrent backcross procedure that may require six to seven sexual generations. However, in wide somatic hybridization it may be possible to rapidly introduce a single chromosome or single gene by regulating somatic growth conditions. This somatic technique will become more useful as efficient cell selection methods that permit identification of agriculturally useful traits are developed.

3. Future prospects of somatic hybridization. Somatic hybridization permits the recovery of new combinations of parental genes to produce unique hybrid plants, such as plants containing the nucleus from one parent and the cytoplasm from the other, or plants containing chloroplast genes from one parent and mitochondria from the second parent. Such cybrids are difficult or impossible to obtain using sexual hybridization. It is now well known that traits controlling herbicide resistance, cytoplasmic male sterility (CMS), disease resistance, response to toxin, and antibiotic resistance are encoded in cytoplasmic organelles (Siddiqui 1990).

Introduction of new cytoplasmic markers will aid in the use of protoplast fusion for transfer of cytoplasm between wild species. For example, genes controlling cytoplasmic male sterility (CMS) can be transferred from one species or variety to a second variety within one cycle of hybridization, in contrast to the lengthy recurrent backcrosses necessary to transfer CMS using conventional methods.

6.4.4 Anther culture

Anther culture is a means to produce plants with a gametic number of chromosomes by aseptic culture of anthers (Fig. 6.22). Figure 6.22 shows callus on 2,4–D medium (left) and young haploid seedlings on medium without 2,4–D (right). The technique of anther culture gives rise to haploid plants either directly or through formation of a haploid callus (Figs. 6.22-1, 6.23). Haploid production has become an integral part of modern plant breeding (Scowcroft 1977, Wenzel 1986, Snape 1986, Bajaj 1990). The technique quickens the breeding cycle by achieving homozygosity in one step through diploidization of the genetic chromosome number. Isolated pollen culture is an extension of anther culture and ensures the production of plants from gametic tissues.

Some examples of doubled haploid systems are given in Tables 6.6 and 6.7. New cultivars obtained by anther culture are constantly being reported from China in crops of economic importance such as rice, wheat, and tobacco. In 1987, the bread wheat variety ‘Florin’ was released in France. This is an awned semidwarf variety developed by the double haploid method (Bajaj 1990).

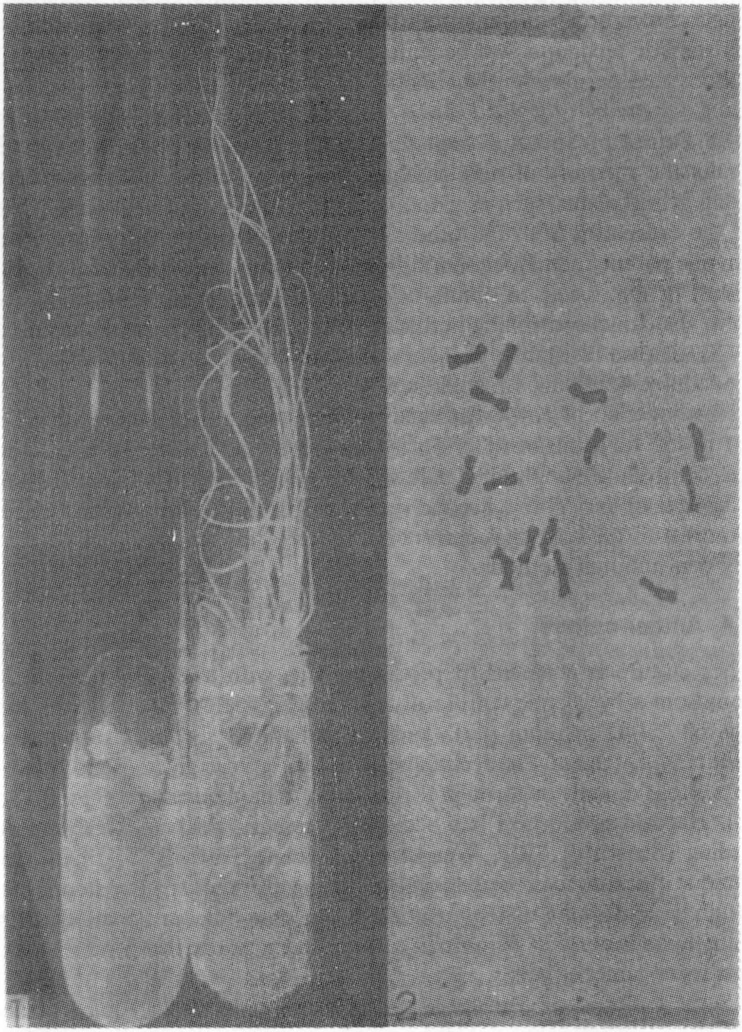

Figure 6.22 Production of haploid plants of *Aegilops* by anther culture of synthetic amphiploid (CC C^uC^u) of *Caudata* and *Umbellulata* (Courtesy: Mikio Kimata and Sadal Sakamoto, 1972).

6.4.5 *In vitro* culture of ovules

In vitro culture of ovules is an important technique in modern plant breeding (Bajaj 1990) and has the following advantages: (a) It is much easier to culture the whole ovules than to isolate single globular embryos, especially

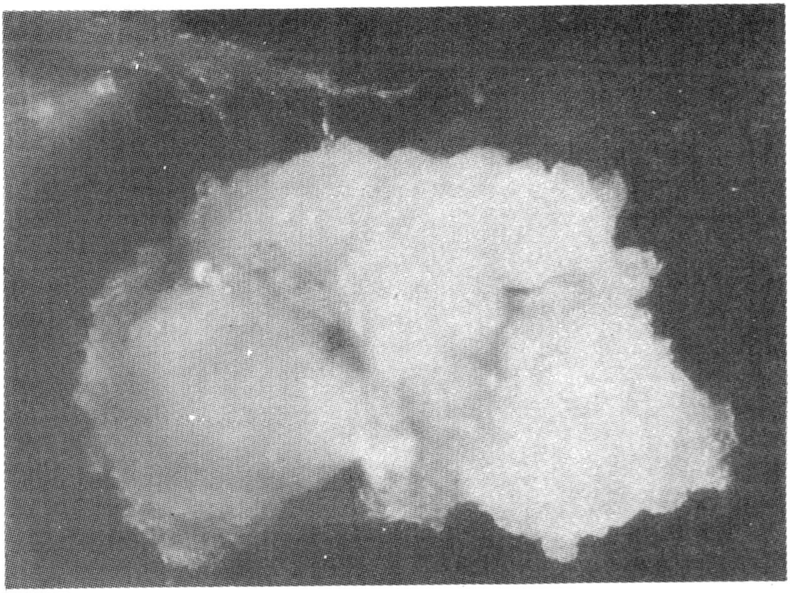

23-a 1st stage

23-b 2nd stage

Figure 6.23-a,b,c,d Stages of haploid production in wheat (Courtesy: Fazal Ali, J. A.).

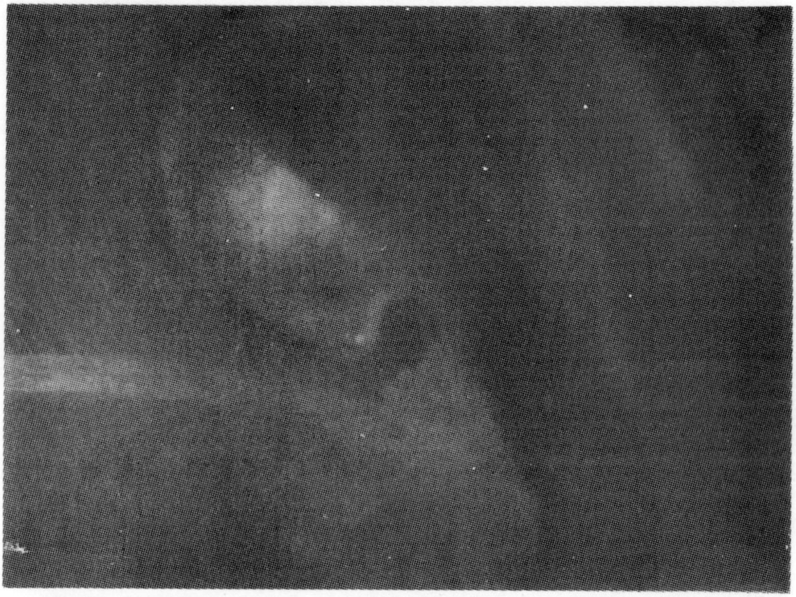

23-c 3rd stage

23-d 4th stage

in small-seeded species. (b) Globular embryos are less damaged when ovules are transferred into the media, as they are not removed from their natural environment. (c) Globular embryos can still be supplied with endogenous and exogenous nutrients. (d) The transfer of ovules enables the culture of hybrid globular embryos at early stages of embryogenesis, that is, before they start to degenerate.

Ovule culture with globular embryos is particularly important in interspecific and intergeneric crosses. Ovule culture (Zenkteler and Nitzsche 1990) allows direct hybridization of various species such as *Glycine*, *Gossypium*, *Impatiens*, *Lycopersicon*, *Nicotiana*, *Brassica*, and *Lolium* × *Festuca*. Several protocols for *in vitro* culture of ovules have been published. One of the most improved protocols for ovule culture has been reported recently by Zenkteler and Nitzsche (1990) for bread wheat.

6.4.6 *In vitro* pollination and fertilization

Zenkteler and Slusarkiewicz–Jarzina (1986) have described *in vitro* pollination and fertilization (Fig. 6.24) in crop plants. The method makes it possible: (i) to study in detail the processes of fertilization and embryogenesis in controlled conditions; (ii) to obtain viable seeds capable of germinating *in situ*; (iii) to raise interspecific and intergeneric hybrid embryos and plants in cases when normal *in vitro* hybridization fails; (iv) to induce parthenogenesis and the development of haploid embryos and plants; (v) to interfere in the development of the female gametophyte following pollination of immature ovules.

In vitro self-pollination of placentas with ovules has been performed successfully with about 35 species, usually those whose ovaries contained a large number of ovules. The process of sexual *in vitro* hybridization has been successfully applied to obtain hybrid embryos and plants among several species of Caryophyllaceae and Solanaceae. In the case of monocotyledons it was possible to obtain well-developed embryos after direct pollination of ovules of *Zea mays* with *Z. mexicana* (Zenkteler and Slusarkiewicz–Jarzina 1986).

6.4.7 Embryo rescue

Plant breeders have been using the embryo rescue technique since 1925. In recent years the technique has been refined (Lange 1986). For cereals, Lange has successfully adopted the following procedure of embryo rescue.

Three to four weeks after pollination the young seeds are surface sterilized. The embryos are isolated and placed on a solid medium containing nutrients and vitamins according to Murashige and Skoog (1962), in half-strength, 3% sucrose and 0.8% agar. The embryos are cultured at 25°C, first in the dark until the seedlings are about 2 cm long and root formation has

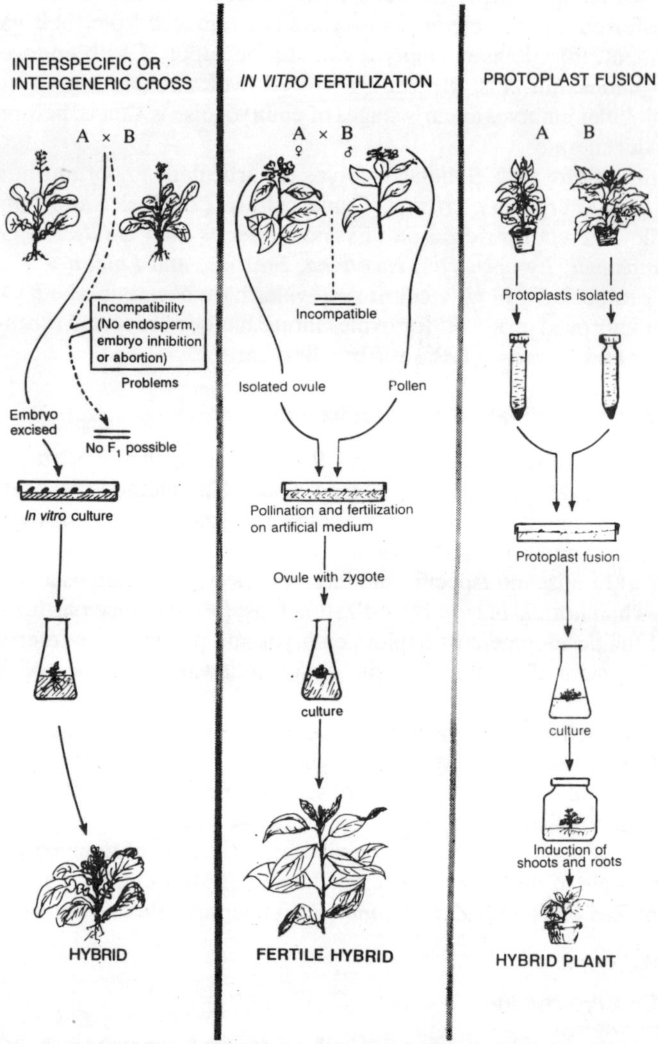

Figure 6.24 Schematic diagram for wide crossing–embryo culture, *in vitro* fertilization, and protoplast fusion. Drawn by Habib-ul-Rahman Mian.

started, and then in the light until the seedlings can be planted in soil. A recent procedure applicable to dicots has been described by Zenkteler and Słusarkiewicz–Jarzina (1986).

Utilization of endomorphic traits in modern plant breeding. Natural and induced variation for endomorphic traits has opened new avenues for future plant breeding (Siddiqui and Yousufzai 1988). It has been suggested by Siddiqui (1990, 1991) that endomorphic traits can be utilized in the construction of new ideotypes for future ecological niches. The traits can also be utilized for breeding new cultivars endowed with enhanced photosynthetic efficiency and nitrogen fixation, and with improved adaptation to high temperature, waterlogging, and salinity.

6.4.8 Micropropagation

Micropropagation may be defined as *in vitro* asexual propagation of crop plants. This technique has the following advantages over the conventional practices of asexual propagation. (a) Only a small amount of plant tissue is needed, as the initial explant is sufficient for regeneration of millions of clonal plants in one year. (b) Micropropagation provides a possible alternative method of propagation for those species which are highly resistant to conventional bulk propagation practices. (c) Micropropagation helps in speedy international exchange of plant material. (d) *In vitro* propagation is not season-dependent (Hu and Wang 1983).

Micropropagation is used for rapid multiplication of elite stocks, elimination of disease, germ plasm preservation, physiological studies, and induction of mutations.

In vitro multiplication can be achieved by (a) enhancing axillary bud breaking, (b) production of adventitious buds, and (c) somatic embryogenesis. In the development of adventitious buds, there is an interplay between the inoculum, the medium, and the culture conditions (Thorpe and Patel 1984). In general, the process of plantlet formation via organogenesis requires induction of shoot buds, development and multiplication of these buds, rooting of the shoots, and hardening of the plantlets.

1. Achievements through micropropagation. Micropropagation has been very successfully employed with different plant species for rapid multiplication and disease eradication. Many of the herbaceous horticultural species are currently propagated commercially using *in vitro* procedures. Examples of successful herbaceous ornamental species propagated through *in vitro* axillary bud proliferation include: *Anigozanthos*, *Anthurium*, *Cephalotus*, *Chrysanthemum*, *Dianthus*, *Fuchsia*, *Gerbera*, *Gladiolus*, *Gloxinia*, *Gypsophila*, *Hosta*, *Phlox*, and *Sacifragra*. Herbaceous vegetables include: *Allium*, *Arachis*, *Asparagus*, *Beta*, *Brassica*, *Cicer*, *Gynara*, *Phaseolus*, *Rheum*, and *Fragaria*. Micropropagation is also successful in the genera *Capsella*, *Glycine*, *Lolium*, *Stellaria*, *Steria*, *Vigna*, and *Zea* (Hu and Wang 1983).

Rigorous attempts have been made to commercially propagate important forest species, but there has been very little success in this area.

Successful examples of axillary bud proliferation include species in the following genera: *Araum*, *Coria*, *Cryptomeria*, *Eucalyptus*, *Populus*, *Tectona*, *Thuja*, and *Sassafras*.

In vitro culture has become the only effective technique for obtaining virus-free plants. Wang and Hu (1980) have reported the regeneration of virus-free plants in 49 species. Also, the African cassava mosaic and cassava brown streak diseases have been eliminated from the East African cassava (*Manihot utilissimus*) cultivars (Kaiser and Teemba 1979). Similarly, rye grass mosaic has been eliminated from *Lolium multiflorum* (Dale 1977) and cocksfoot streak virus, mild mosaic virus, and mottle virus from *Dactylis glomerata* (Dale 1979). Various viral diseases of Citrus spp. (Navarro et al. 1975), *Pisum sativum* (Kartha and Gamborg 1978), and *Trifolium repens* (Bernett et al. 1975) have also been controlled via micropropagation. This technique has also been applied very successfully in regenerating virus-free apple plants (Huang and Millikan 1980). In Pakistan, virus-free potatoes have been produced and multiplied through micropropagation.

2. Future prospects. *In vitro* micropropagation is an extremely time and space-saving technique. In the future this technique will become the main method for cloning desirable genotypes of forest, ornamental, and fruit trees. Rapid progress in micropropagation can be made if optimal temperature, photoperiod, light intensity, and light quality levels are identified for each target species.

6.4.9 *In vitro* germ plasm preservation

New advances are constantly taking place in the preservation of plant germ plasm. Due to continued depletion of naturally occurring reservoirs of germ plasm of agriculturally important plant species, global efforts are being made to resort to non-conventional methods, such as *in vitro* preservation methods. Such methods are particularly useful in some plants, not only for the induction of new genetic variability, but also for the long-term conservation of germ plasm for the future. For this purpose, plant tissue culture combined with cryogenic methods is an optimistic approach (Bajaj 1986). *In vitro* germ plasm preservation is particularly useful in the situations described by Nitzsche (1983).

Protocol for germ plasm preservation. The *in vitro* preservation of plant material can be achieved in two ways: (a) slow growth (Withers 1985), and (b) cryopreservation.

CRYOPRESERVATION. The technology of freeze preservation of cells and meristem culture is gaining considerable momentum (Bajaj 1986, 1990). The basic principle of cryopreservation involves subjecting a culture to super-low temperatures in the presence of cryoprotectants in order to bring the cells to a state of non-dividing and inactive metabolism. Work in cryobiology

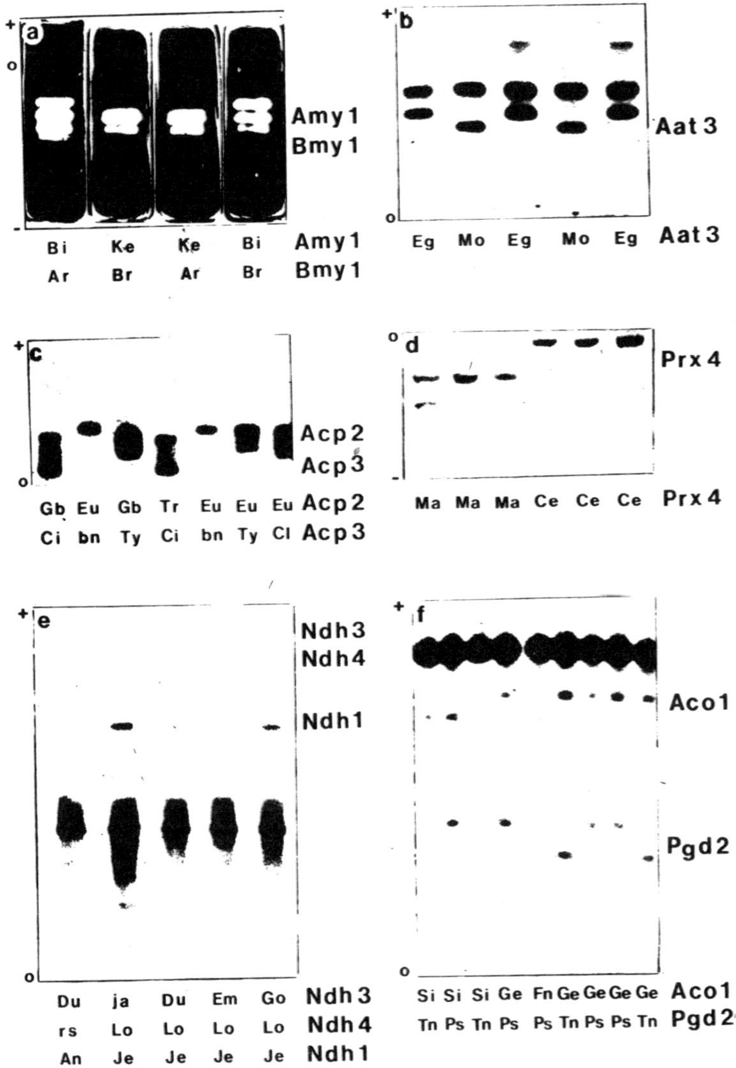

Figure 6.25 Isoenzyme loci in barley (Courtesy: I.B. Linde-Laursen, Gunnar Nielsen, Hanne Bay Johanson, 1987, Risö, Denmark).

during the last 10 years has shown that entire plants can be regenerated from cells, embryos, and meristems frozen and stored in liquid nitrogen (–196°C). This has been done with a wide range of plant species, including tobacco, carrot, atropa, petunia, carnation, potato, cassava, strawberry, sugarcane, rice, oil palm, populus, and some legumes. These results have created optimism about the feasibility of using cryogenic methods as a tool in modern plant breeding (Nitzsche 1983, Bajaj 1986).

6.4.10 New biochemical and molecular inputs in plant breeding

1. Isozymes in plant breeding. During the 1950's, considerable evidence was already available indicating that plant enzymes exist in multiple forms (Siddiqui et al. 1972, Chevre et al. 1991). During the last 40 years, isozyme research has gained considerable momentum. The contributions of isozyme loci to plant breeding have been recently reviewed (Siddiqui 1991). In maize, perhaps more than in any other plant species, there has been a broad range of studies involving isozymes (Goodman and Stuber 1983). These have varied from purely biochemical studies of enzymatic properties to plant breeding studies concerned with the maintenance of quality control during seed production. In addition, most of the structural genes encoding maize isozymes have been localized to specific chromosomal sites. Certain isozymes, most notably alcohol dehydrogenase (ADH), have been studied intensively. Products of two loci, Adh 1 and Adh 2, have now been sequenced. The technique has also been very useful in barley (Fig. 6.25). (a) of Figure 6.25 shows α- and β-amylase; (b) aspartate aminotransferase; (c) acid phosphatase; (d) peroxidase; (e) NADH dehydrogenase; and (f) aconitate hydratase and phospho-gluconate dehydrogenase.

2. New advances in molecular biology. The rapidity with which the new inputs of molecular biology have been delivered to the doorstep of the modern plant breeder is remarkable. The first restriction enzyme that cuts DNA at specific sites was discovered in 1970, and by 1972 the first recombinant DNA molecules had been generated. The next year, genetically engineered plasmids had been functionally cloned in a bacterium. In 1977, recombinant DNA technology went commercial, eukaryotic genes were found to contain noncoding regions (introns), and procedures for rapid DNA sequencing became available.

Since then new developments have been constantly taking place in plant molecular biology (Siddiqui 1991, Karpinski et al. 1994). Repeated nucleotides have several attributes that make them potentially superior to chromosome pairing or polymorphism in protein or allozyme loci for phylogenetic inferences in higher plants. Because each sequence family comprises many individual members, polymorphism in interspecific comparisons is more meaningful than polymorphism at single-copy loci (Siddiqui 1991).

3. New class of genetic markers in plant breeding. Restriction fragment length polymorphisms (RFLP's) are receiving particular attention as a new class of genetic markers, and are being used extensively in genetic engineering research (Figdore 1988, May 1990). There are two forms of repeated sequence polymorphism (Siddiqui 1991). One is polymorphism in the abundance of a sequence (RSAP), and the other is polymorphism in the length of restriction fragments (RFLP).

There are several advantages of RFLP markers over other types of genetic markers (May 1990). While morphological and allozyme markers are limited in number and placement throughout the genome, RFLP markers are potentially unlimited due to the large number of fragments produced by restriction digestion and due to the number of different enzymes available for identification of RFLP's (Henry and de Buyser 1990). Unlike many morphological markers, exact genotypic information can be obtained when evaluating RFLP's. Since neither environmental effects nor epistatic effects affect the interpretation of RFLP loci, these markers can be scored simultaneously with ease even in single plants. Genotypes at RFLP loci can be scored at any stage of development as well as in any plant part. Genotypic evaluation requires only small amounts of plant tissue, not enough to be totally destructive to the plant being sampled (May 1990).

There are four basic applications of RFLP markers in plant breeding programmes (Lydiate et al. 1993).

a. Individual RFLP's can be used to mark a single gene of interest.

b. Many RFLP's dispersed throughout the genome can be used to genetically characterize individuals or populations.

c. RFLP's can be used to analyze and dissect quantitative traits.

d. RFLP's can be used to select for quantitative trait loci (QTL's) based on association of QTL's with linked RFLP markers (Figdore 1988, May 1990).

RFLP's (restriction fragment length polymorphisms) have been particularly helpful in elucidating phylogenetic relationships among wild and cultivated species of crop plants (Siddiqui 1991). RFLP is indeed a very efficient tool in the hands of modern plant breeders, and will be increasingly utilized to solve the problems of contemporary and future plant breeding (Lydiate et al. 1993).

RAPD's (random amplified polymorphic DNA's) are new genetic markers available to modern plant breeding. These are polymerase chain reaction (PCR)-based markers, which detect the primer binding sites of intervening DNA (Lydiate et al. 1993).

TABLES

Table 6.1 Selected examples of utilization of induced mutations through evolution of improved varieties of *Triticum aestivum*

Improved trait	Variety	Country	Year of release
High yield	NP 836	India	1961
	Stadler	USA	1964
	E mai 6	China	1966
	Zenkouzi–Momugi	Japan	1969
	Yuang Nong 53	China	1970
	Yuang Nong 61	China	1971
	Nyan Chung 7112	China	1974
	Odesskaja 75	USSR	1975
	Nan Jing 3	China	1976
	Yuan Feng 4	China	1978
	Jauhar 78	Pakistan	1979
	Emai No. 9	China	1980
	Kijanka	USSR	1981
	Yuanfu No. 2	China	1982
	352	China	1983
	Longfumai No. 1	China	1984
	Longfumai No. 2	China	1985
	Xifu 5	China	1985
	Spinnaker	Italy	1987
	Chuanfu 3	China	1989
	Soghat 90	Pakistan	1991
High biomass	Heichun No. 2	China	1979
Reduced height			
	Lu Mai No. 6	China	1984
	Tambo	Switzerland	1985
	Lu Mai 8	China	1985
Early maturity	NP 836	India	1961
	Stadler	USA	1964
	Tai Fu 1	China	1966
	Tai Fu No. 15	China	1968
	Zenkouzi–Komugi	Japan	1969
	Yuang Nong 53	China	1970
	NI 5643	India	1975
	Jing Fen 1	China	1977
	Yuan Dong 7848	China	1978
	Yu Yuan 1	China	1979
	Yuan Fuzao	China	1980

Improved trait	Variety	Country	Year of release
	Chuanfu No. 1	China	1982
	Lu Mai No. 4	China	1983
	Yuandong 94	China	1984
	Longfumai No. 2	China	1985
	Chuanfu 2	China	1989
Lodging resistance	Els	FRG	1960
	Stadler	USA	1964
	092	China	1966
	Yuan Feng 1	China	1968
	Sirius	FRG	1969
	Novosibirskaja 67	USSR	1969–70
	Xin Shu Kuang 1	China	1971
	Yuan Chung 7112	China	1974
	Odesskaja 75	USSR	1975
	Nan Jing 3	China	1976
	Jing Fen 1	China	1977
	MY 8	Hungary	1978
	Yuan Feng 4	China	1978
	Yu Yuan 1	China	1979
	Yuanfeng 5	China	1983
	Spinnaker	Italy	1987
	Wanyuan 75–6	China	1979
	Lu Mai No. 4	China	1983
	Xinchun No. 2	China	1984
	Longfumai No. 2	China	1985
Winter hardiness	Stadler	USA	1964
	1161	China	1966
	Yuan Feng 1	China	1968
	Zhonga No.1	China	1969
	Yuan Feng 3	China	1971
	Jiaxuan No.1	China	1974
	503	China	1975
	Yuan Dong 772	China	1977
	Changwei 51503	China	1983
Sprouting resistance	Hankkijas Taava	Finland	1978
Awning	NP 836	India	1961
Tillering	Taifu No. 22	China	1968
	Yuan Chung 7112	China	1974
	503	China	1975
	Yuan Dong 772	China	1977
	Changwei 51503	China	1983

Improved trait	Variety	Country	Year of release
Efficient fertilizer use	Shan Nong Radio 63	China	1980
Good photosynthetic ability	Shan Nong Radio 63	China	1980
Drought resistance	NP 836	India	1961
	Tai Fu 1	China	1966
	Fai Fu No.22	China	1968
	Jien Mai 2	China	1969
	Qichun No.1	China	1971
	Wuchun No.3	China	1973
	Fuer	China	1977
	Yu Yua	China	1979
	Chuanfu No.1	China	1982
	Changwei 51503	China	1983
	Xinchun No.2	China	1984
	Zlatostrui	Bulgaria	1985
	Longfumai No.3	China	1986
Shattering resistance	IAS 63	Brazil	1974
	Jauhar 78	Pakistan	1979
Disease resistance	NP 836	India	1961
Leaf rust	Sinvalocho Gama	Argentina	1962
	Stadler	USA	1964
	E Mai 6	China	1966
	Yuang Nong 53	China	1970
	Xin Shu Kuang 1	China	1971
	Kexing No.15	China	1972
	Ning Mai 3	China	1973
	Yuan Chung 7112	China	1974
	503	China	1975
	Xifu No.3	China	1977
	Yuan Dong 7848	China	1978
	Altimir 67	Bulgaria	1979
	Yun Fuzao	China	1980
	Payne	USA	1981
	Yuanfeng 5	China	1983
	Yuandong 94	China	1984
	Yuandong 83	China	1985
	Wei Fu 6757	China	1986
	Soghat 90	Pakistan	1991
Stem rust	Sinvalocho Gama	Argentina	1962
	Stadler	USA	1964
	E Mai 6	China	1966
	Yuang Nong 53	China	1970

Improved trait	Variety	Country	Year of release
	Yuang Nong 61	China	1971
	Kexing No.15	China	1972
	Ning Mai 3	China	1973
	Yuang Chung 7112	China	1974
	N 1 5643	India	1975
	Fuer	China	1977
	Neimai No.5	China	1979
	77 L 15	China	1883
	Yuandong 94	China	1984
	Lu Mai 8	China	1985
	Longfumai	China	1986
Stripe rust	Stadler	USA	1964
	1161	China	1966
	Yuang Feng 1	China	1968
	Jein Mai 2	China	1969
	Yuang Nong 53	China	1970
	Yuang Nong 61	China	1971
	Kexing No. 15	China	1972
	Jiaxuan No.1	China	1974
	7112	China	1975
	Xifu No.3	China	1977
	Yuan Dong 772	China	1978
	Yuan Dong 7848	China	1979
	Yuandon No.2	China	1982
	Yuanfeng	China	1983
	Lu Mai No.5	China	1984
	Yuandong 83	China	1985
	Wei Fu 6757	China	1986
	Chuanfu 2	China	1989
Powdery mildew	Kexing No.15	China	1972
	Ning Mai 3	China	1973
	Jiaxuan No.1	China	1974
	Fuer	China	1977
	Chang Wei	China	1978
	Altimir 67	Bulgaria	1979
	Emai No.9	China	1980
	Lu Mai 5	China	1084
	Zlatostrui	Bulgaria	1985
Scab tolerance	Zhemai No. 3	China	1983
	Longfumai No. 2	China	1985
Baking quality	Els	FRG	1960
	Sirius	FRG	1969

Improved trait	Variety	Country	Year of release
	Novosibirskaja 67	USSR	1969–70
	Qichun No. 1	China	1971
Amber grain colour	Sharbati Sonora	India	1967
	Zhonga No. 1	China	1969
	Pusa Lerma	India	1971
	NI 5643	India	1975
	Jauhar 78	Pakistan	1979
Nutritive value	Sharbati Sonora	India	1967
Chapati-making quality	Sharbati Sonora	India	1967
	Jauhar 78	Pakistan	1979
	Yuandong 94	China	1984
	Longfumai No. 2	China	1985
White grain colour	Yuan Feng 3	China	1971
	NI 5643	India	1975
	Yuan Dong 772	China	1977
	Yuan Dong 7848	China	1978
Soft grain	Stadler	USA	1964
High grain weight	Qichun No. 1	China	1971
	Kexing No. 15	China	1972
	Humai No. 3	China	1978
	Shan Nong Radio 63	China	1980
Elasticity	Pusa Lerma	India	1971
High gluten	Pusa Lerma	India	1971
Bold lustrous grain	NI 5643	India	1975
Large grain	Lu Ten 1	China	1968
	Xin Shu Kaung 1	China	1971
	Wuchun No. 3	China	1973
High protein content	Humai No. 3	China	
	Carolina	Chile	
	Yuanfeng 5	China	
	Longfumai No. 1	China	
Wide adaptability	Taifu No. 10	China	1968
	Yuandon No. 2	China	1982
	Xifu 4	China	1985
	Wei Fu 6757	China	1986
Heat tolerance	Neimai 5	China	1979
	Emai No. 9	China	1980
	Lu Mai No. 5	China	1984
	Yuandong 83	China	1985
Salt tolerance	Jiaxuan No. 1	China	1974
	Emai No. 9	China	1980
	Changwei 510003	China	1983

Improved trait	Variety	Country	Year of release
	Wei Fu 6757	China	1986

Source: Siddiqui (1991); Mutation Breeding Newsletters 27(1986); 30(1987); 32(1988); 37(1991).

Table 6.2 Selected examples of utilization of induced mutations through evolution of improved varieties of *Oryza sativa*

Name of variety	Country	Year of release	Main improved attributes of variety
SH 30–21	China	1957	Higher yield, short growing period.
YH 1	China	1963	Higher yield, short growing period.
Fu She 31	China	1965	Early maturity, short culm, resistant to stress, suitable for mountain areas.
Ai Fu No. 9	China	1966	Short culm, blast resistance, higher yield.
Er Fuzao	China	1968	Early maturity.
Jin Fu No. 8	China	1969	Early maturity, shorter culm, *Xanthomonas* resistance.
Iratom 38	Pakistan	1970	3–4 weeks earlier and 2% higher in protein than mother variety.
PARC 1	Philippines	1970	Narrower and longer grains with less chalky areas.
PARC 2	Philippines	1970	5–10 days earlier in maturity, slightly narrower and much longer grains with less chalky areas, good eating quality, high acceptance by farmers.
BPI–121–407	Philippines	1971	Early, short, stiff-strawed, high-tillering, high-yielding, good disease resistance, moderately resistant to bacterial leaf blight.
Fu Xuan 124	China	1972	Blast resistance, intermediate maturity.
IIT 48	India	1972	One week earlier, finer grain, better test weight compared with IR8.
Nucleoryza	Hungary	1972	Earliness, with maintained blast resistance and improved yield.
Hybrid mutant 95	India	1973	Short stature, short duration, high yield potential, photoperiod and thermo-insensitive, wide adaptability, high protein (12.3%) and high lysine (4.07%) content, covers more than 20,000 ha in the Punjab, very popular for transplanting late in the season and for multiple cropping schemes.

Name of variety	Country	Year of release	Main improved attributes of variety
Fu 709	China	1974	Higher yield, cold resistant.
Wan Geng 257	China	1975	Fertilizer tolerance, blast resistance, higher yield.
Calrose 76	USA	1976	Short stature (ca 96 cm against 120 cm of Calrose) due to shortening of all internodes, otherwise essentially unchanged.
Xiang Fudao	China	1976	Cold tolerance in seedling period, blast and *Xanthomonas* resistant.
Akihikari	Japan	1977	Season-neutral ripening within 100 days, most suitable in tight crop rotations.
Kashmir Basmati	Pakistan	1977	Early maturity, suitable to be grown in Azad Kashmir at 2000–5000 feet altitude, 30% higher yield than coarse varieties grown there.
Fu Wan 23	China	1978	Yellow stunt and *Xanthomonas* resistant, bigger spike and large grain, good quality.
Wan Fu 33	China	1978	Early maturity, cold resistance, blast tolerance.
Callendal	France	1979	Longer grain, improved threshability, better tolerance to *Sclerotium oryzae*.
Calmochi 201	USA	1979	Glutinous endsoperm.
CNM 25	India	1979	Earlier maturing 14–25 days, increased effective tillering, long grain; 15% higher yield per day, resistant to thrips.
CNM 31	India	1979	Earlier-maturing 10 days, increased effective tillering, long grain, 9% higher yield per day, resistant to BLB, BLS, BPH, brown spot.
Fu Zhu	China	1979	Early maturity, cold tolerance, blast resistance, lodging resistance.
Guang Dabai	China	1979	Intermediate maturity, higher yield.
CNM 6	India	1980	Early maturing 15–23 days, increased effective tillering, long grain, 10% higher yield per day, drought tolerant.
CNM 20	India	1980	Earlier-maturing 10–12 days, increased effective tillering, long grain, resistant to BLC, BLS, BPH.
She Fu 802	China	1980	Early maturity, high yield, disease resistance.

Name of variety	Country	Year of release	Main improved attributes of variety
Calmochi 202	USA	1981	Semidwarf type, glutinous (waxy) grain, awnless.
M114	China	1981	Cold tolerance, Fulgorid resistance.
M–302	USA	1981	Semidwarf type, intermediate maturity, medium grain shape, more lodging resistant, higher yielding than M–301.
Schuang Ke No. 1	China	1981	Intermediate maturity, higher yield, higher temperature tolerance.
Atinuta–1	Indonesia	1982	Early maturity, improved resistance to brown plant hopper and green leafhopper, blast, high tolerance to salinity, high protein content, good eating quality.
Wanhua	China	1983	Semidwarf, disease-resistant, superior grain quality, high yield 1986:100,000 ha.
Qingwei No. 1	China	1985	High-yielding, disease-resistant late season variety.
Shadab	Pakistan	1987	High-yielding, improved grain quality.
Valencia 87	USA	1987	Lodging-resistant, high yield up to 12 t/ha.
Radiation 9–1	China	1988	115–120 day tillering, yield ca. 7.5 t, earlier and 9% higher-yielding than popular var. Guanglui 4, wide spectrum of resistance to BLB 1986–89: ca. 29,000 ha.
Radiation 85–63	China	1989	Higher-tillering, higher spikelet fertility; good cooking quality; good adaptation to mountainous districts.

Table 6.3 Examples of utilization of induced mutations through evolution of improved varieties of *Gossypium* spp.

Name of variety	Country	Year of release	Main improved attributes of variety
MCU 7	India	1971	Early maturity (10–15 days), higher yields.
Rasmi	India	1976	Day-length tolerant, suitable for cultivation throughout India during main kharif season and also off-season, high yield, superior quality.
Lu Main No. 1	China	1976	Good vigour and plant architecture, higher boll production.

Name of variety	Country	Year of release	Main improved attributes of variety
Pusa Ageti	India	1978	High ginning GP 38, short duration 150 days, jassid tolerant, fits well for cotton/wheat rotation in Northern India.
NIAB–78	Pakistan	1983	High yield, early maturity, medium stature with 0–2 monopodials.

Table 6.4 Examples of utilization of induced mutations through evolution of improved varieties of leguminous crops

Name of variety	Country	Year	Main improved attributes of the variety
			***Cicer arietinum* L.**
Hyprosola	Bangladesh	1981	Matures 10 days earlier, more pods, higher harvest index, higher planting density feasible, 19% higher yield than parent variety, 4% more protein.
CM 72	Pakistan	1983	Resistant against chickpea blight *Ascochyta rabiei*, high yield.
NIFA–88 = CM–1918	Pakistan	1990	Moderately resistant to *Ascochtya rabiei*, 15–10% more yield, fixes more nitrogen.
			***Vigna radiata* L.**
Pant Moong 2	India	1982	Moderately resistant to mung bean yellow mosaic virus (MYMV), higher yield.
TAP–7	India	1982	5–7 days earlier maturity, fairly tolerant to powdery mildew and leaf spot disease, 23% higher yield than 'Kopergaon'.
Co 4	India	1982	High-yielding, early-maturing, drought resistant.
NIAB Mung–18	Pakistan	1983	Early and uniform maturity, high yield.
MN 121–25	Pakistan	1986	Early, uniform maturity, high yield, greater tolerance to MYMV.
NM 19–19	Pakistan	1986	Early uniform maturity, high yield, greater tolerance to MYMV.
MN 19–19	Pakistan	1986	Early uniform maturity, high yield, greater tolerance to MYMV.

Name of variety	Country	Year	Main improved attributes of the variety
MN 20–21	Pakistan	1986	Earlier, more pods, harvest index 31%, TGW 38.6 g, 65% higher yield than Pak. 22, better tolerant to MYMV.
NM 13–1	Pakistan	1986	Earlier, shorter, more pods, harvest index 28%, 1000-seed weight 40.5 g, 44% higher yield than parent (6601).
NM 51	Pakistan	1990	Early, non-shattering, resistant to MYMV and *Cercospora* leaf spot (CLS), high yield.
MN 54	Pakistan	1990	Early, non-shattering, resistant to MYMV and CLS, high yield.

Table 6.5 Induced somaclonal variation for useful agronomic characters in crop plants

Crop	Explant	Induced somaclonal variation
Wheat	Immature embryo	Spike morphology, plant height, tillering, maturity, leaf waxiness, awnless, gliadins, and *a*-amylase.
Rice	Seed embryo	Panicle size, number of tillers, plant height, days to flower, seed fertility.
Maize	Immature embryo	Seedling mutants, endosperm MtDNA sequence rearrangement, *S. maydis* resistance, and T toxin resistance.
Oat	Apical meristem, immature embryo	Heading date, awns, plant height, leaf striping.
Sugar	Various	Sugar yield, eyespot, culmicolous cane spot diseases, Fiji virus, downy mildew, esterase isozymes, auricle length.
Potato	Protoplast, leaf callus	Tuber yield, resistance to early and late blight, morphology of stem, leaf, flower, tuber, maturity and plant habit.
Rape and mustard	Meristems, anthers, embryo	Glucosinolates, *Phoma lingam* mustard tolerance, waxiness, seed index, growth habit, earliness, pods per plant, field resistance to important diseases.
Tobacco	Anthers, protoplasts, leaf callus	Alkaloids, reducing sugars, plant height, leaf size, specific leaf chlorophyll loci.
Alfalfa	Immature ovaries	Plant height, multifoliate leaves, plant habit, dry matter yield, leaf petiole.

Source: After Scowcroft (1984) and Katiyar and Chopra (1990).

Table 6.6 Some examples of the double haploid system in crop plants

Crop	Method of haploid production	Reference
Rice	Anther culture	Ying and LiLiang–Tsai
Wheat	Anther culture, interspecific hybridization	De Buyser et al.
Barley	Interspecific hybridization, anther culture	Kasha and Kao Freidt and Foroughi–Wehr
Tobacco	Anther culture	Collins and Legg
Cotton	Spontaneous	Tarcotte and Feaster

Source: Snape 1986.

Table 6.7 A summary of experiments on the reproducible regeneration of crop plant microspores

Crop	Year	Workers
Brassica campestris	1975	Keller, Rajhathy, Lacapra
	1979	Keller, Armstrong
Brassica napus	1975	Thomas, Wenzel
	1981	Lichter
Ipomoea batatas	1978	Sehgal
Oryza perennis	1975	Oono
Secale cereale	1975	Thomas, Wenzel
	1982	Foroughi–Wehr, Friedt, Wenzel
Triticale	1973	Sun, Wang, Chu
Zea mays	1975	Anonymous
	1979	Hu Han, Hao Shui

Source: Adapted from Wenzel (1986).

STUDY QUESTIONS

1. Explain mutation breeding with reference to the improvement of important agricultural crops.
2. Name the different mutagenic agents. In what way(s) can their effects be modified?
3. Outline a generalized scheme of mutation breeding. What steps are involved in the evolution of a mutant variety?
4. Discuss the role of plant cytogenetics in modern plant breeding.
5. Write short notes on the following: (a) chromosome mapping in rice; (b) telomeric sequences in rye; (c) synthesis of multigeneric hybrids without embryo culture; (d) induction of polyploidy.

6. Discuss the techniques and advantages of protoplast culture.
7. *In vitro* techniques are assuming increasing importance in contemporary plant breeding. Discuss this issue.
8. Explain what is meant by *micropropagation*. What are the present achievements and future prospects of this technique?
9. Discuss some new biochemical and molecular inputs in plant breeding.
10. Explain each of the following: (a) cryopreservation, (b) RFLP, (c) anther culture, (d) nucleo-cytoplasmic heterosis, (e) embryo rescue translocation, (f) point mutation.

REFERENCES

Aslam, R. M.; Siddiqui, K. A. Cytogenetic analysis of induced variability in *Pennisetum americanum*. In: Siddiqui, K. A.; Faruqui, A. M. eds. 1986:p. 289–296.

Bajaj, Y. P. S. Prospects of the *in vitro* induction and cryopreservation of genetic variability in crop improvement. In: Siddiqui, K. A.; Faruqui, A. M. eds. 1986:p. 53–54.

Bajaj, Y. P. S. ed. Biotechnology in agriculture and forestry, Vol. 13. Wheat. Berlin. Springer Verlag; 1990

Bajaj, Y. P. S. Biotechnology in wheat breeding. In: Bajaj, Y. P. S. ed. 1990:p. 2–23.

Bell, G. D. H. Investigations in *Triticinae*. I. Colchicine techniques for chromosome doubling in interspecific and intergeneric hybridization. J. Agric. Sci. 40:9–18; 1950.

Bernett, O. W.; Gibson, P. B.; Sepo, A. A comparison of heat treatment and meristem tip culture for obtaining virus free plants of *Trifolium repens*. Plant Dis. Rep. 54:834–837; 1975.

Blakeslee, A.; Avery, A. Methods of inducing doubling of chromosomes in plants. J. Hered. 28:393–411; 1937.

Carlbom, C. Depolyploidization, evolution and improvement of hexaploid bread wheat (*Triticum aestivum* L. em.) ssp. *vulgare* (Vill. Host, Mackey). Hereditas 62:437–443; 1969.

Cauderon, Y.; Saigne, B. Obtention d'amphidiploides and partir de F_1 (*Triticum aestivum x Secale cereale*). Ann. Amelior. Plant. 11:369–373; 1961.

Chevre, A. M.; This, P.; Eber, F.; Deschamps, M.; Ranard, M. Delseny, M; Quiros, C. F. Characterization of disomic addition lines *Brassica napus – Brassica nigra* by isozyme, fatty acid, and RFLP markers. Theoretical and Applied Genetics 81:43–49; 1991.

Cocking, E. C. Plant cell protoplasts isolation and development. Ann. Rev. Plant Physiol. 23:29–50; 1972.

Collins, G. B.; Legg, P. D. Recent advances in the genetic applications of haploidy in *Nicotiana*. In: Davies, P. R.; Hopwood, D. A. eds. The plant genome. Norwich, Connecticut: The John Innes Charity; 1981:p. 197–213.

Cowen, N. M.; Thompson, S. A.; Wilkinson, T. C. Culture associated variation in maize inbreds. Plant Breed. 104 (2):134–143; 1990.

Dale, P. J. The elimination of rye-grass mosaic virus from *Lolium multiflorum* by meristem tip culture. Ann. Ap. Biol. 85:93–96; 1977.

Dale, P. J. The elimination of cocksfoot streak virus, cocksfoot mild mosaic virus and cocksfoot mottle virus from *Dactylis glomerata* by shoot tip and tiller bud culture. Ann. Ap. Biol. 93:285–288; 1979.

De Buyser, J.; Henry, Y.; Laur, R.; Lonnet, P. Utilisation de l'angrogenese *in vitro* dans des programmes de selection du ble tendre (*Triticum aestivum* L.) Z. Pflanzenzuecht 87:290–99; 1981.

Dix, P. J.; Street, H. E. Sodium chloride resistant cultured cell lines from *Nicotiana sylrestris* and *Capsicum annuum*. Plant Sci. Lett. 5:231–237; 1975.

Driscoll, C. J. The contribution of cytogenetics to plant improvement. In: Chopra, V. L.; Joshi, B. C.; Sharma, R. P.; Bansal, H. C. eds. Genetics: New frontiers, Vol. IV. New Delhi: Oxford and IBH Publishing Co.; 1984:p. 69–76.

Elliott, F. C. Plant breeding and cytogenetics. New York: McGraw–Hill Book Company; 1958.

Evans, D. A. Protoplast fusion. In: Evans, D. A.; Sharp, W. R.; Ammirato, P. V.; Yamada, Y. eds. Handbook of plant cell culture, Vol. I. London: Macmillan Publishing Co.; 1983: p. 299–321.

Evans, W. D. Improving vegetatively propagated polyploids using induced polyploids of interspecific hybrids. In: Siddiqui, K. A.; Faruqui, A. M. eds. 1986:p. 377–385.

Figdore, S. S. Applications of RFLPs to problems in plant breeding. In: Zakri, A. M. ed. Plant breeding and genetic engineering. Malaysia: SABRAO; 1988:p. 361–376.

Fladung, M. Transformation of diploid and tetraploid potato clones with rol C gene of *Agrobacterium rhizogenes* and characterization of transgenic plants. Plant Breed. 104(4):295–304; 1990.

Freidt, W.; Foroughi–Wehr, B. Anther culture of barley (*Hordeum vulgare* L.). Plant regeneration and agronomic performance of homozygous diploid progenies. In: Asher, M. J. C. et al. eds. Proc. Fourth Int. Barley Genet. Symp. Edinburgh: Edinburgh University Press; 1981: p. 690–698.

Frey, K. J. Mutation breeding for quantitative attributes. In: The use of induced mutations in plant breeding. Radiat. Bot. Suppl. 5:465–475; 1965.

Fromm, M. E.; Taylor, L. P.; Walbot, V. Stable transformation of maize after gene transfer by electroporation. Nature 319:791–793; 1986.

Gale, M. D.; Youssefian, S. Dwarfing genes in wheat. In: Russell, G. E. ed. Plant breeding. London: Butterworths; 1985:p. 1–35.

Goodman, M. M.; Stuber, C. W. Maize. In: Tanksley, S. D.; Orton, T. J. eds. Isozymes in plant genetics and breeding, Part B. Amsterdam: Elsevier Science Publisher B.V.; 1983: p. 1–33.

Gregory, W. C. Mutation frequency, magnitude of change and the probability of improvement in adaptation. In: Use of induced mutations in plant breeding. Rad. Bot Suppl. 5:429–441; 1965.

Gustafson, J. P.; Kaltsikes, P. J. The effect of rye heterochromatin on the stability and performance of triticale. In: Siddiqui, K. A.; Faruqui, A. M. eds. 1986:p. 345.

Haberlandt, G. Kulture versuche mit isolierten Pflanzenzellen. Mat. Naturwiss kl. Abt. 1. 111:69–92. Wien: Kais. Akad. Wiss.; 1902.

Hagberg, A. Production of duplications in barley breeding. Hereditas 48:243–246; 1962.

Hagberg, P.; Hagberg, A. Segmental interchanges in barley, III. Translocations involving chromosomes 6 and 7 used in production of duplications. Z. Pflanzenzuecht. 81:111–117; 1978.

Henry, Y.; de Buyser, I. Wheat anther culture: Agronomic performance of doubled haploid lines and the release of a new variety 'Florin'. In: Bajaj, Y. P. S. ed. 1990:p. 285–352.

Heyser, J. W.; Nabors, M. W. Osmotic adjustment of tobacco cells and plants to penetrating and non-penetrating solutes. Plant Physiol. Suppl. 63:77; 1979.

Hu, C. Y.; Wang, P. J. Meristem, shoot tip and bud cultures. In: Evans, D. A.; Sharp, W. R.; Ammirato, P. V.; Yamada, Y. eds. Handbook of plant cell cultures, Vol. I. London: Macmillan; 1983:p. 177–227.

Huang, S. C.; Millikan, D. F. *In vitro* micro grafting of apple shoot tips. HortSci. 15:741–743; 1980.

IAEA. Manual on mutation breeding. Technical Series No. 119. Joint FAO/IAEA Division, under the auspices of the International Atomic Energy Agency. Vienna: IAEA 1977.

Junker, R.; Zimney, J.; Luhrs, R.; Lorz, H. Transient expression of chimaeric genes in dividing and non-dividing cereal protoplasts after PEG-induced DNA uptake. Plant Cell Rep. 6:329–332; 1987.

Kaiser, W. J.; Teemba, L. R. Use of tissue culture and thermotherapy to free East African cassava cultivars of African cassava mosaic and cassava brown streak disease. Plant Dis. Rep. 63:780–784; 1979.

Karpinski, S.; Karpinska, B.; Wingsle, G.; Hällgren, J. E. Molecular responses to phyto-oxidative stress in *Pinus sylvestris*. I. Differential expression of nuclear and plastid genes in relation to recovery from winter stress. Physiol. Plant 90:358–366; 1994.

Kartha, K. K.; Gamborg, O. L. Meristem culture techniques in the production of disease free plants and freeze preservation of germplasm of tropical tuber crops and grain legumes. In: Maraite, H.; Meyer, J. A. eds. Diseases of tropical food crops. Louvain (Belgium): Universite Catholique de Louvain; 1978:p. 267– 283.

Kasha, K. L.; Kao, K. N. High frequency haploid production in barley (*Hordeum vulgare* L.). Nature 225:874–876; 1970.

Katiyar, R. K.; Chopra, V. L. Somaclonally induced earliness in a *Brassica juncea* germplasm accession with field resistance to important diseases. Plant Breed. 104 (3):262–264; 1990.

Kerber, E. R. Wheat: reconstitution of the tetraploid component (AABB) of hexaploids. Science 143(3603):253–255; 1964.

Kihara, H. Importance of cytoplasm in plant genetics. In: Siddiqui, K. A.; Faruqui, A. M. eds. 1986:p. 419–430;

Krikorian A. D.; Berquam, D. L. Plant cell and tissue cultures. The role of Haberlandt. Bot. Rev. 35:59–88; 1969.

Krishnamurthi, M. Notes on disease resistance of tissue culture sub clones and fusion of sugarcane protoplasts. ISSCT Sugarcane Breed. Newsl. 35:24–26; 1974.

Lange, W. Reconstitution of hexaploid wheat from *Triticum dicoccoides* (AABB, 2n = 28) and *T. tauschii* (DD, 2n = 14). In: Horn, W.; Jensen, C. J.; Odenbach, W.; Schieder, O. eds. Genetic manipulation in plant breeding. Berlin: Walter de Gruyter; 1986:p. 175–177.

Larkin, P. J.; Scowcroft, W. R. Eye spot disease of sugarcane: Host specific toxin induction and its interaction with leaf callus. Plant Physiol. 67:408–414; 1981.

Law, C. N.; Snape, J. W.; Worland, A. J. Quantitative genetic studies in wheat. In: Sakamoto, Sadao. ed. Proc. 6th Intl. Wheat Genetics Symposium. Kyoto (Japan): Faculty of Agriculture, Kyoto University; 1983:p. 539–547.

Lee, B. T.; Murdoch, J.; Topping, J.; Kreis, M; Jones, M. G. K. Transient gene expression in wheat (*Triticum aestivum*) protoplasts. In: Bajaj, Y. P. S. ed. 1990:p. 659–668.

Liu, M. C.; Chen, W. H. Tissuc and cell culture as aids to sugarcane breeding, 1. Creation of genetic variability through callus culture. Euphyt. 25:393–403; 1978.

Liu, M. C.; Chen, W. H. Application of tissue and cell culture techniques for sugarcane improvement. Annual Report Research Development Council, Taiwan Sugar Corp., Taiwan. pp. 14–15; 1982.

Lydiate, D.; Sharpe, A.; Lagercrantz, U.; Parkin, I. Mapping the *Brassica* genome. Outlook on Agriculture 22:85–92; 1993.

Maan, S. S.; Gordon, J. Compendium of alloplasmic lines and amphiploids in the *Triticeae*. In: Miller, T. E.; Koebner, R. M. D. eds. 1988:p. 1325–1349.

McIntosh, R. A. Catalogue of gene symbols for wheat. In: Miller, T.E.; Koebner, R. M. D. eds. 1988:p. 1225–1323.

MacKey, J. Radiogenetics in *Triticum*. Genet. Agr. XII. Fasc. 3–4:210–230; 1960.

McVittie, M. A.; Gale, M. D.; Marshall, G. A.; Westcott, B. The intrachromosomal mapping of the Norin 10 and Tom Thumb dwarfing genes. Heredity 40:67–70; 1978.

May, C. E. Triticale x wheat hybrids. In: Bajaj, Y, P. S. ed. 1990:p. 229–249.

Miller, T.E.; Koebner, R. M. D. eds. Proc. 7th International Wheat Genetics Symposium, Cambridge England. Trumpington, Cambridge, England: Institute of Plant Science Research, Cambridge Laboratory; 1988

Mochizuki, A.; Shigenaga, S. Extraction of B genome from hexaploid wheat. Science. Rep. Hyogo Univ. Agric. 6–2 Series: Agric. Biol.:49–52; 1964.

Morris, R. Chromosomal locations of genes for wheat characters. Annual Wheat Newsletter 26:14–27; 1980.

Murashige, T.; Skoog, F. A revised medium for rapid growth and bioassays with tobacco tissue cultures. Physiol. Plant. 15:473–497; 1962.

Nabors, M. W.; Gills, S. E.; Bernstem, C. S.; Mais, M. E. NaCl tolerant tobacco plants from cultured cells. Z. Pflanzenphysiol. 97:13–17; 1980.

Navarro, L.; Poistacher, C. N.; Murashige, T. Improvement of shoot tip grafting *in vitro* for virus free citrus. J. Am. Soc. Hort. Sci. 100:471–479; 1975.

Nitzsche, W. Germplasm preservation. In: Evans, D. A.; Sharp, W. R.; Ammirato, P. V.; Yamada, Y. eds. Handbook of plant cell culture, Vol. I. London: Macmillan; 1983:p. 782–805.

Ou–Lee, T. M.; Turgeon, R.; Wu, R. Expression of a foreign gene linked to either a plant-virus or a *Drosophila* promoter, after electroporation of protoplasts of rice, wheat and sorghum. Proc. Natl Acad. Sci., USA, 83:6815–6819; 1986.

Perez de la Vega, M. Biochemical characterization of populations. In: Hayward, M. D.; Bosemark, N. O.; Ramagosa, I. Plant breeding – principles and prospects. London: Chapman and Hall: 1993:p. 184–200.

Power, J. B.; Chapman, J. V. Isolation, culture and genetic manipulation of plant protoplasts. In: Dixon, R. A. ed. Plant cell culture: a practical approach. Oxford: IRL Press; 1985:p. 37–66.

Ramage, R. T. Chromosome aberrations and their use in genetics and breeding translocations. Barley Genet. 1:99–117; 1964.

Ranade, S.; Feierabend, J. Comparison of light induced stress reactions in susceptible and paraquat tolerant green cell cultures of *Chenopodium rubrum* L. J. Plant Physiol. 137 (6):749–752; 1991.

Rick, C. M. The potential of exotic germplasm for tomato improvement. In: Vasil, I. K.; Scowcroft, W. R.; Freyeds, K. J. Plant improvement and somatic cell genetics. New York: Academic Press; 1982:p. 1–28.

Saeed Iqbal, R. M.; Chaudhry, M. B.; Aslam. M.; Bandesha, A. A. Economic and agricultural impact of mutation breeding in cotton in Pakistan. Proc. Contribution of Plant Mutation Breeding to Crop Improvement. Vienna, Austria: FAO/IAEA; 1991:p. 187–201.

Scowcroft, W. R. Prospects for crop improvement through plant tissue culture. In: Muhammad, A.; Aksel, R.; Von Borstel, R. C. eds. Genetic diversity in plants. New York: Plenum Press; 1977:p. 417–430.

Scowcroft, W. R. Somaclonal variation—a 'new' genetic resource. In: Chopra, V. L.; Joshi, B. C.; Sharma, R. P.; Bansal, H. C. eds. Genetics: New frontiers. New Delhi: Oxford, IBH Publishing Co.; 1984:p. 35–48.

Sears, E. R. Amphidiploids in seven-chromosome *Triticeae*. Res. Bull. MO Agric. Exp. Sta. 336:66; 1941.

Sears, E. R. Wheat cytogenetics. Ann. Rev. Gen. 3:451– 468; 1969.

Sears, E. R. An induced mutant with homologous pairing in wheat. Can. J. Genet. Cytol. 19:585–593; 1977.

Shepherd, K. W.; Islam, A. K. M. R. Fourth compendium of wheat–alien chromosome lines. In: Miller, T. E.; Koebner, R. M. D. eds. 1988:p. 1373–1395. .

Shigenaga, S. Genome deletion in wheat. In: Siddiqui, K. A.; Faruqui, A. M. eds. 1986:p. 361–371.

Shuaib, M.; Siddiqui, K. A.; Salahuddin, M.; Hoda, S. Improvement of cotton through induced mutations and hybridization. Nucl. 18:37–42; 1981.

Siddiqui, K. A. General discussion on aneuploidy in wheat genetics. Proc. 2nd Int. Wheat. Genetics Symp. Lund, Sweden. Hereditas Suppl. (2):392; 1963.

Siddiqui, K. A. Cytogenetic studies in the tribe Triticeae. Reading (England): University of Reading. Ph.D. Thesis. 1964. 434 p.

Siddiqui, K. A. Genetic analysis of bread wheat. Nucl. (Pakistan). 6:146–150; 1969.

Siddiqui, K. A. Induction of polyploidy in multigeneric hybrids in the group Triticeae. Hereditas 67:191–204; 1971a.

Siddiqui, K. A. Extraction of AADD component of *Triticum aestivum*. Hereditas 68:151–158; 1971b.

Siddiqui, K. A. Partitioning of an allopolyploid. Hereditas 71:289–300; 1972a.

Siddiqui, K. A. An evaluation of wheat mutants for useful agronomic characters. Nucl. 9:29–35; 1972b.

Siddiqui, K. A. Protein content and quality of wheat chromosome substitution lines. Hereditas 71:157–160; 1972c.

Siddiqui, K. A. Resistance in some wild and cultivated grasses to the phytotoxicity of a systemic fungicide. Exper. 28:1239–1240; 1972d.

Siddiqui, K. A. Synthetic amphiploids in breeding-genetics and evolutionary studies in wheat. In: Muhammad, A.; Aksel, R.; Von Borstel, R. C. eds. Genetic diversity in plants. New York: Plenum Press; 1977:p. 97–102.

Siddiqui, K. A. Induced amphiploidy in crop plants. In: Singh, R. B.; Singh, R. M.; Singh, B. D. eds. Advances in cytogenetics and crop improvement. New Delhi: Kalyani Publishers; 1982a:p. 107–117.

Siddiqui, K. A. Realities and possibilities of crop improvement. Pakistan Agriculture 4(8):22–24; 1982b.

Siddiqui, K. A. Utilization of induced variations in wheat improvement. In: Sakamoto, Sadao ed. Proc. 6th Intl. Wheat Genetics Symp., Kyoto (Japan): Plant Germ-plasm Institute, Faculty of Agriculture, Kyoto University; 1983:p. 79–80.

Siddiqui, K. A. Mutations in wheat-future possibilities. In: Bajaj, Y. P. S. ed. 1990:p. 549–578.

Siddiqui, K. A. Phylogenetic inferences from cytogenetic and molecular mechanisms in the tribe Triticeae. In: National Seminar on the Role of Plant Health and Care in Agriculture, Univ. of Agriculture, Faisalabad. Proceedings volume. Faisalabad: University of Agriculture; 1991:p. 50–60.

Siddiqui, K. A.; Arain, M. A. Induced variation for pelshenke value and caryopsis weight in bread wheat. Cereal Res. Commun. 14(4):379–385; 1986.

Siddiqui, K. A.; Arain, M. A.; Jafri, K. A. Evolution of high yielding wheat variety through fast neutron treatment. Mut. Breed. N.L. 18:6–7; 1981.

Siddiqui, K. A.; Arain, M. A.; Jafri, K. A. Use of kernel weight as criterion for the selection of productive mutants in wheat. Rachis 6(2):53; 1987.

Siddiqui, K. A.; Arain, M. A.; Jafri, K. A. Use of wheat mutants in conventional breeding programmes. In: Strategies for stabilizing wheat production in the Punjab. Lahore: Department of Agriculture, Government of the Punjab; 1988:p. 31–38.

Siddiqui, K. A.; Doll, H. Screening for improved protein quality mutants in wheat. Z. Pflanzenzuecht. 70:143–147; 1973.

Siddiqui, K. A.; Faruqui, A. M. eds. New genetical approaches to crop improvement. Tandojam: Atomic Energy Agricultural Research Centre; 1986.

Siddiqui, K. A.; Haahr, V. Different reactions of wheat mutants to a systemic fungicide. Naturwiss. 58:415–416; 1971.

Siddiqui, K. A.; Ingversen, J.; Koie, B. Inheritance of protein patterns in a synthetic allopolyploid to *Triticum monococcum* (AA) and *Aegilops ventricosa* (DDMM). Hereditas 72:205–214; 1972.

Siddiqui, K. A.; Jafri, K. A. Tracing the course of mutations in an allopolyploid. In: Siddiqui, K. A.; Faruqui, A. M. eds. 1986:p. 257–288.

Siddiqui, K. A.; Jafri, K. A.; Arain, M. A. Sodium azide mutagenesis in wheat: mutants with golden glumes. Mut. Breed. N.L. 33:19; 1989.

Siddiqui, K. A.; Jones, J. K. Cytogenetic studies of intergeneric hybrids involving *Aegilops ventricosa* and species of *Triticum*. Can. J. Genet. Cytol. 9:776–784; 1967a.

Siddiqui, K. A.; Jones, J. K. The D genomes of *Aegilops crassa* Boiss. Bol. Genet. 4:29–31; 1967b.

Siddiqui, K. A.; Jones, J. K. Chromosome pairing in a pentaploid hybrid involving *Triticum durum*, *Aegilops squarrosa* and *Aegilops bicornis*. CARYAB 21:339–342; 1968a.

Siddiqui, K. A.; Jones, J. K. The crossability of varieties of tetraploid rye with some varieties and monosomic lines of *Triticum aestivum*. Sindh University Sci. Res. J. 3:165–168; 1968b.

Siddiqui, K. A.; Jones, J. K. Genetic necrosis in *Triticum* × *Aegilops* pentaploid hybrids. Euphyt. 18:71–78; 1969.

Siddiqui, K. A.; Khan, M. I.; Arain, M. A.; Jafri, K. A. Genetic analysis of protein content, lysine content and related characters in *Triticum aestivum*. In: Seed protein improvement by nuclear techniques. Vienna: International Atomic Energy Agency; 1978a: p. 455–466.

Siddiqui, K. A.; Mustafa, G.; Arain, M. A.; Jafri, K. A. Realities and possibilities of improving cereal crops through mutation breeding. Proc. Contribution of Plant Mutation Breeding to Crop Improvement. Vienna, Austria: FAO/IAEA; 1991:p. 173–185.

Siddiqui, K. A.; Rajput, M. A.; Arain, M. A.; Arain, A. Ghafoor; Jafri, K. A. Wheat improvement by induced mutations. 18th Gamma Field Symp., Ohmiya–Machi Japan; Naka, Ibaraki (Japan): Institute of Radiation Breeding NIAS MAFF; 1979:p. 33–47.

Siddiqui, K. A.; Rajput, M. A.; Arain, M. A.; Jafri, K. A. Induced variability for useful agronomic traits in bread wheat. Ramanujam, S. ed. Proc. 5th Intl Wheat Genetics Symposium, New Delhi. New Delhi: Secretary, Indian Society of Genetics and Plant Breeding, I.A.R.I.; 1978b:p. 547–558.

Siddiqui, K. A.; Rajput, M. A.; Arain, U. A. Induced seed protein mutants of *Triticum aestivum*. Naturwiss. 62:393; 1975.

Siddiqui, K. A.; Yousufzai, M. N. Natural and induced variation for endomorphic traits in the tribe *Triticeae*. In: Miller, T. E.; Koebner, R. M. D. eds. 1988:p. 139–143.

Siddiqui, N. H. Studies on colchicine induced autotetraploid *Solanum americanum* Mill. and its triploid and aneuploid progenies. In: Siddiqui, K. A.; Faruqui, A. M. eds. 1986:p. 387–394.

Sigurbjornsson, B. Induced mutations. In: Crop breeding. Madison, Wisconsin: American Society of Agronomy and Crop Science Society of America; 1983:p. 153–176.

Singh, R. M.; Singh, J. P. Singh, J. Influence of seed pre-soaking on mutagenic efficiency of EMS in wheat (*Triticul aestivum*). In: Siddiqui, K. A.; Faruqui, A. M. eds. 1986:p. 97–107.

Snape, J. W. The use of doubled haploids in breeding self pollinating crops. In: Siddiqui, K. A.; Faruqui, A. M. eds. 1986:p. 27–34.

Stadler, L. J. Mutation in barley induced by X-rays and radium. Science 68:186–187; 1928a.

Stadler, L. J. Genetic effects of X-rays in maize. Proc. Natl. Acad. Sci. USA 14:69–75; 1928b.

Tarcotte, E. L.; Feaster, C. V. Methods of producing haploids: semigametic production of cotton haploids. In: Kasha, K. J. ed. Haploids in higher plants, advances and potential. Guelph, Canada: University of Guelph; 1974: p. 53–64.

Thorpe, I. A.; Patel. K. R. Clonal propagation: adventitious buds. In: Vasil, I. K. ed. Cell culture and somatic cell genetics of plants, Vol. 1. New York: Academic Press; 1984:p. 49–60.

Tsuchiya, T. Chromosome mapping in barley by means of trisomic analysis. In: Siddiqui, K. A.; Faruqui, A. M. eds. 1986:p. 325–344.

Tsunewaki, K. Genetic diversity of the cytoplasm in *Triticum* and *Aegilops*. Tokyo: Jap. Soc. Promot. Sci.; 1980:p. 290.

Wang, P. J.; Hu, C. Y. Regeneration of virus free plants through *in vitro* culture. In: Fitcher, A. ed. Advances in biochemical engineering, plant cell cultures II, Vol 18. Berlin: Springer Verlag; 1980:p. 61–99..

Wenzel, G. Cell genetics and haploids in crop improvement. In: Siddiqui, K. A.; Faruqui, A. M. eds. 1986:p. 1–14.

Withers, L. A. Cryopreservation and storage of germplasm. In: Dixon, R. A. ed. Plant cell culture: A practical approach. Oxford: IRL Press; 1985:p. 169–199.

Ying, C.; Li Liang–Tsai. Investigation and utilization of pollen derived plants in rice and wheat. In: Davies, P. R.; Hopwood, D. A. eds. The plant genome. Norwich, Connecticut: The John Innes Charity; 1980.

Zenkteler, M.; Nitzsche. W. *In vitro* culture of wheat ovules. In: Bajaj, Y. P. S. ed. 1990:p. 250–261.

Zenkteler, M.; Slusarkiewicz–Jarzina, A. Sexual reproduction in plants by applying the method of test tube fertilization of ovules. In: Horn, W.; Jensen, C. J.; Odenbach, W.; Schieder, O. eds. Genetic manipulation in plant breeding. Berlin: Walter de Gruyter; 1986:p. 415–423.

Outline of Chapter 7

7. EXPERIMENTAL DESIGN AND ANALYSIS

Syed Sadaqat Mehdi[1]
Iftikhar Ahmad Khan

LEARNING OBJECTIVES

After studying this chapter, you should be able to:

- Understand and apply the fundamentals of statistics to methods of plant breeding.
- Plan, analyse, and interpret the results of an experiment.
- Estimate heritability.
- Estimate genetic advance.

7.1 Statistics

Statistics is an important tool for all types of research. Statistical principles are independent of subject matter, and statistical procedures are used successfully in an ever-increasing number of disciplines including agriculture and biological sciences. The validity of experimental results depends on employing appropriate experimental design and accurate data collection methods. The major objective of this chapter is to provide the basic statistical techniques useful for a student of plant breeding.

7.1.1 Definition of statistics

Statistics deals with the methods of collecting, organizing, summarizing, presenting, and analysing data, drawing valid conclusions, and making reasonable decisions on the basis of such analysis. Statistical data can be obtained by conducting suitable laboratory or field experiments, or by organizing well designed surveys. Therefore, statistics contributes to scientific research by providing the tools to analyse data gathered through experimentation and observation.

---

[1]Dr. Syed Sadaqat Mehdi and Dr. Iftikhar Ahmad Khan are with the Department of Plant Breeding and Genetics, University of Agriculture, Faisalabad.

7.1.2 Population, sample, and variables

Two concepts concerning the aggregate of observations are **population** and **sample**. In the statistical sense a **population** consists of all possible values (measurements, counts) of the character under study. The simplest examples of a population are the two faces of a coin (head, tail) or the six faces of a die. The set of all plants of a given F_2 cross between two wheat varieties with plant height from 60–90 cm also constitutes a population.

Ideally, an entire population should be observed. However, the researcher is usually not able to study the whole population, because it is too expensive and time-consuming to take all possible measurements. Therefore, one deals with a part of the population. This portion of a population is called a **sample**. A sample is a subset of items drawn from the population under study. It is taken to be representative of the larger population. Statistics is based on the idea that the sample will be 'typical' in some way and that it will enable us to make predictions about the whole population. The faces marked 3, 5, 6 on a die constitute a sample of three from a population of six, i.e. {1, 2, 3, 4, 5, 6}.

For a sample to be useful, it should be a true representative of the population. One way to ensure this is to draw the sample at random. A **random sample** is drawn in such a way that every individual of the population has an equal chance of being selected. The manner of selecting or including the individuals should not favour any individual against another. This is best done through the use of tables of random numbers.

A population can be **finite** (having an end or limit) or **infinite** (having no end). The population is generally hypothetical, while a sample is something that is actually handled. Any quantitative measure that describes a characteristic of a population is called a **parameter**. Parameters are the constants that are peculiar to a particular population. They are rarely precisely obtained but are estimated. A quantitative measure that describes a characteristic of a sample is called a **statistic**. For example, the height of the tallest plant in the F_2 population of a wheat cross is a parameter. The height of the tallest plant in a sample of 300 plants from the same population is a statistic.

Statistical information is expressed in different ways. The characteristic that is being studied, such as the height of individuals or hair colour, is called a **variable**. A variable is often represented as a symbol, such as x, y, a, b, etc., which can assume any of a prescribed set of values, which is called the **domain** of the variable. If the variable can have only one value it is known as a **constant**. Variables are of two types, qualitative and quantitative. A **qualitative variable** is identified simply by noting its presence. It has no numerical value, e.g. the outcome of tossing a coin is either head or tail. If a variable consists of numerical values it is called a **quantitative variable**, e.g. the number of bacteria in a culture or the height of an individual plant expressed in centimetres.

Quantitative variables are further classified as **continuous** and **discrete/discontinuous.** A variable which can theoretically assume any value between two given values is called a **continuous variable.** It can assume any numerical value over a single interval or over different intervals; thus the observations can be measured to any degree of accuracy on a numerical scale. A **discrete variable** is one whose possible values consist of certain points between successive values. The number of children in a family is an example of a discrete variable since the only values that the variable can take are whole number values such as 0, 1, 2, 3 and so forth; a family cannot have 2.5 children.

Data which can be described by discrete or continuous variables are called discrete or continuous data, respectively. Discrete data can only take a sequence of distinct values, and usually consist of counted observations or a series of integers, whereas continuous data are observations which can take any value on a continuous scale.

7.2 Measures of variability

The measurement and evaluation of variability is extremely important in agricultural and biological sciences. It is essential in order to draw meaningful conclusions from a given set of observations. There are two important measures of the closeness of the data in a set to a common centre: measures of central tendency and measures of dispersion.

7.2.1 Measures of central tendency

A measure of central tendency or centrality is a number which is somewhere near the middle of the data values. It indicates a centre of the distribution which the data represent. The commonly used measures of central tendency are (i) arithmetic mean, (ii) median, and (iii) mode.

Arithmetic mean. The **mean** is one of the most important and commonly used measures of central tendency. The arithmetic mean is an ordinary average and is the sum of all observations divided by the total number of observations.

Suppose there are *n* observations with numerical values denoted as x_1, x_2, ..., x_n. When the variable values are denoted by x, the symbol for the arithmetic mean is $\bar{x}$ (read: 'x bar'). The arithmetic mean ($\bar{x}$) is given as:

$$\bar{x} = (x_1 + x_2 + ... + x_n)/n = (\Sigma x)/n$$

where $x_1, x_2, ..., x_n$ are the *n* different values of the character observed. The symbol Σ is the uppercase Greek letter 'sigma', which stands for the summation. Similarly, if the data were to be denoted by $y_1, y_2, ..., y_n$ then the arithmetic mean would be denoted by $\bar{y}$.

Applying the above formula, the arithmetic mean of the set of values (93, 97, 88, 98, 95) which represent the height of five wheat plants in centimetres is calculated as:

$$\bar{x} = \frac{93+97+99+98+95}{5} = \frac{471}{5} = 94.2 \text{ cm}$$

The arithmetic mean is affected by every item in the sample but disproportionately so by extreme values. It has both advantages and disadvantages. The advantages include: (a) it is the most commonly used; (b) it is easy to compute; and (c) it is easy to understand. The one major disadvantage is that it is unduly affected by extreme values and may therefore be not truly representative of the sample.

Median. One commonly used alternative method for finding a central value is the **median**, which is the middle value when the data are arranged in order from smallest to largest. This indicates that generally half of the data will be larger than the median and half will be smaller. The sample median is denoted by *M*, and is important as a measure of position or location.

If the number of data values is odd, then the median is simply the middle number. The previous data set 93, 97, 88, 98, 95 has the median 95, because 95 is the third observation (middle number) when the data are arranged in ascending order as 88, 93, 95, 97, 98. If the number of data values is even, then the median is the average of the two middle numbers. For example, the median of the observations 85, 97, 89, 87, 91, and 98 is 90, as it is the average of 89 and 91, the third and four observations, when the data are arranged as 85, 87, 89, 91, 97, 98.

Mode. The **mode** is the value that occurs most frequently in the data, and is denoted by *MO*. Of all the measures of central tendency, the mode is easiest to obtain; it can be determined by inspection once the data are arranged. For example, the data set 7, 11, 7, 5, 13 has mode 7. A distribution having only one mode is called **unimodal**. If a data set has more than one mode, it is **multimodal**.

7.2.2 Measures of dispersion

The variability of the data is also referred to as its **spread, dispersion**, or **scatter**. The extent to which data values differ from each other overall is called **dispersion**. For example, examine the following two samples.

Sample 1: 12.3, 13.2, 12.0, 12.9, 13.8
Sample 2: 17.4, 6.9, 6.0, 19.8, 14.1

In the first sample, the values range from 12.0 to 13.8, while in the second sample the range is from 6.0 to 19.8. Thus the variability in sample 2 is greater than that in sample 1, although the means of the two samples are the same, each being 12.84 The mean does not tell us anything about the spread of the values. Therefore, common measures of dispersion such as **range, mean deviation, variance**, and **standard deviation** are used.

Range. The simplest measure of dispersion is **range.** This is the difference between the largest and the smallest values in the data. Because it is computed in this way, the range can never be a negative number. The range of the data set 15, 9, 15, 6, 25, 36, 30, 15 is (36 − 6) = 30. The range is very easy to compute, but is often considered inefficient because it concentrates entirely on the extreme values.

Mean deviation or average deviation. The mean deviation or average deviation is also called **mean absolute deviation.** In contrast with the range, the average deviation takes into account the magnitude of all the observations in the data. It is defined as the average of the absolute values of the differences between each data value and the mean. The absolute value of a number is the number without the associated sign and is indicated by two vertical lines placed around the number. Thus $|-7| = 7$; $|+5| = 5$; $|-0.35| = 0.35$.

The mean deviation is calculated by adding the absolute values of the deviations of individual observations from their arithmetic mean and dividing the sum by the number of observations. The formula for mean deviation is:

$$\text{Mean deviation (M. D.)} = \frac{\Sigma|x-\bar{x}|}{n}$$

where $\bar{x}$ is the arithmetic mean of observations, and $|x - \bar{x}|$ is the absolute value of the deviation of x from $\bar{x}$. The mean deviation of the data set 4, 5, 9, 12, 15 is calculated as:

$$\text{Arithmetic mean} = \bar{x} = \frac{4+5+9+12+15}{5} = 9$$

$$\text{Mean deviation} = \frac{|4-9|+|5-9|+|9-9|+|12-9|+|15-9|}{5}$$

$$= \frac{|-5|+|-4|+|0|+|3|+|6|}{5}$$

$$= \frac{5+4+0+3+6}{5} = 3.6$$

Variance. The most commonly used, and generally most useful, measure of dispersion is the **variance.** This is based upon an average of the squared differences between each data value and the mean. The **variance** of a data set is defined as the arithmetic mean of the squares of the deviations of the individual values from their arithmetic mean. Thus, if we have *n* values $x_1, x_2, ..., x_n$ in the data, then

$$\text{variance} = \frac{(x_1 - \bar{x})^2 + (x_2 - \bar{x})^2\ ) + . . . + (x_n - \bar{x})^2}{n}$$

$$= \frac{\Sigma(x_i - \bar{x})^2}{n}$$

where: i stands for all individual values in turn, i.e. $x_1, x_2, x_3 \ldots x_n$
n = the total number of individual values

The variance is represented by two symbols: σ^2 (the Greek letter sigma; read as 'sigma square' or 'sigma squared') stands for a population, and S^2 ('S square' or 'S squared') for a sample. σ^2 is a parameter which is constant for a particular population, whereas S^2 is a statistic which changes from sample to sample from the same population. Variance is expressed in the units of the observation, squared, as for example cm^2 (centimetres squared). The variance must be positive since it represents the arithmetic mean of positive quantities. The magnitude of the variance clearly reflects the scatter in the data; a larger variance reflects larger deviations.

A divisor of *n* is used when defining the population variance (σ^2). However, when defining the sample variance (S^2), $(n-1)$ is used as divisor. $(n-1)$ is referred to as the **degree of freedom** associated with the sample variance. The concept of degrees of freedom is that in any given sample set there is one value that is not truly independent. For example, if we are testing four varieties in four adjoining plots and establish planting order by drawing names from a box, if variety A is planted in the first plot, variety B in the second, and variety C in the third, then variety D *has to* go in the fourth plot; i.e. its position is already determined. The use of $(n-1)$ is said to give an unbiased estimator of σ^2 since not all members of the population could be included.

Symbolically, the sample variance (S^2) for observations $x_1, x_2, ..., x_n$, is given by:

$$S^2 = \frac{(x_1 - \bar{x})^2 + (x_2 - \bar{x}) + \ldots + (x_n - \bar{x})^2}{n - 1}$$

$$= \frac{\Sigma (x_i - \bar{x})^2}{(n - 1)}$$

The numerator of S^2 is often referred to as the **sum of squares** and is denoted by SS. For example, the sum of squares for the data set 4, 8, 10, 14, is:

$$(4-9)^2 + (8-9)^2 + (10-9)^2 + (14-9)^2$$
$$= (-5)^2 + (-1)^2 + (+1)^2 + (+5)^2 = 52$$

and the sample variance is 52/3 = 17.33

The sample variance can be expressed in either form (a) or (b):

$$\text{a)} \quad S^2 = \frac{\left[\sum x^2 - \frac{(\Sigma x)^2}{n}\right]}{(n - 1)}$$

$$\text{b)} \quad S^2 = \frac{(\Sigma x^2 - n\bar{x}^2)}{(n - 1)}$$

Standard deviation. The most useful measure of dispersion is the **standard deviation**. This is defined as the square root of the variance. The standard deviation is expressed in the same units of measurement as the observation, and for this reason it is preferred to the variance as a descriptive measure. However, it is often easier to work with variances. Thus the two measures, variance and standard deviation, are complementary. The standard deviation computed from a sample of data is often called the sample standard deviation or the standard deviation of the sample. Standard deviations for populations and for samples are represented by σ and S, respectively. Hence σ is a parameter and S is a statistic. Thus the sample standard deviation is:

$$S = \sqrt{S^2}$$

The sample standard deviation of the previous data set 4, 8, 10, 14 is $\sqrt{17.33} = 4.16$.

Standard deviation of the mean. The **standard deviation of the mean** is one of the most useful statistics and is often called the **standard error**, or less frequently, **standard error of the mean**. Usually the term standard deviation applies to observations and standard error applies to means. The standard error is inversely proportional to the square root of the number of observations in the mean and is calculated either as in (a) or (b):

$$\text{a) } sx = \frac{s}{\sqrt{n}}$$

$$\text{b) } sx = \sqrt{\frac{s^2}{n}}$$

For example, the data set 4, 8, 10, 14 has $s\bar{x} = s/\sqrt{n} = 4.16/\sqrt{4} = 2.08$.

Coefficient of variation. The **coefficient of variation** (CV) is a measure of variability, defined as the ratio of the standard deviation to its mean, expressed as a percentage.

$$CV = s/\bar{x} \times 100$$

For the example, 4, 8, 10, 14 with standard deviation 4.16 and mean 9, the coefficient of variation is

$$CV = (4.16/9) \times 100 = 46.22\%$$

The coefficient of variation is independent of the scale in which the observations are obtained, i.e. thus the CV of a given set of length measurements is the same whether the measurements are made in inches or centimetres. The CV is thus an ideal device for comparing variabilities of characters which have widely differing units of measure, for example in comparing the variability in the height of a set of plants (cm) with the variability of grain weight (mg) of the same set of plants.

In agricultural experiments, the CV is a good index of the reliability of the experiment, as it indicates the degree of precision with which the experiment was carried out. The higher the CV value, the lower the reliability of the experiment. The CV differs greatly with the type of experiment, the crop plant, and plant trait measured. For example, the acceptable range of CV is 6-8% for variety, and 10-12% for fertilizer trials of rice. The CV's for rice plant height and yield are 3% and 10%, respectively.

7.3 Experimental design

An **experiment** is a planned inquiry to obtain new facts or to confirm or deny the results of previous experiments. An experiment is designed to provide answers to specific questions, and its design determines how experimental procedures are carried out. The researcher takes the results from a large number of readings and observes the relationships found in his data. Sometimes the results of experimental inquiry are utilized in making administrative or advisory decisions like recommending a variety or an insecticide.

7.3.1 Objectives of experimental design

Design of an experiment means planning an experiment so that relevant information is collected in a way that will allow the researcher to answer his question. Sound experimental design is necessary to ensure that the assumptions required for appropriate interpretation of the data are satisfied. Therefore, before beginning an experiment the researcher must choose an appropriate design, so that appropriate data will be obtained in a way that permits an objective analysis leading to valid inferences with respect to the stated problem. The accuracy and sensitivity of the results will also be increased.

The purpose of an experimental design is to provide a maximum amount of information relevant to the problem under study at the minimum cost. Before the planning of an experiment, the principles of scientific experimentation such as identification and statement of the problem, formulation of the specific objectives, selection of the experimental material, estimating the number of replications, and selection of the design should be followed. Statistical methods are of great help in planning experiments and in the subsequent analysis of the data. The progress of research in any discipline depends on precise definition of the problems.

Once a problem has been well defined, the next step is to determine the parameters to be measured. These measurements are basic quantities such as weight, length, etc. Good measurements should be accurate and free from bias. All measurements contain a component of error because no two experiments are identical, except in an approximate sense. It is a matter of common experience that the results obtained by repeating scientific experiments show variation from one set to another. There are many reasons why small variations will occur in repeated tests, however, a well planned experiment reduces this variation considerably. If the variation is small the experiment is said to be **repeatable**.

The effect of a set of treatments must be tested by using an experimental design which allows for unbiased estimates of the true treatment differences with a specified degree of precision. Therefore, every effort should be made to adopt a design which gives a low experimental error. There are several useful experimental designs which can be used according to different situations. These include: (a) completely randomized design, (b) randomized complete block design, (c) Latin-square design, (d) lattice design. In this chapter discussion is confined to the first two.

7.3.2 Completely randomized design

This is the simplest design, and the one on which all others are built. In this design, treatments are arranged completely at random with respect to the experimental units, or vice versa. This design imposes no restrictions on the allocation of the treatments to the experimental units. The completely randomized design is usually selected when the overall variation of the experimental material is relatively small. Hence it is considered to be most useful in laboratory techniques, greenhouse studies, or in animal experiments where environmental effects are much alike.

Advantages. The main advantages of the completely randomized design are:

a. The statistical analysis is extremely simple.

b. The design is completely flexible in that any reasonable number of treatments and replications may be used. Moreover, the treatments need not be repeated the same number of times.

c. Missing observations do not present any difficulty in the statistical analysis.

d. For small experiments, a more precise estimate of error is obtained, especially for those experiments where degrees of freedom for experimental error are less than 20. Therefore, for a given number of treatments and replications, the design allows for the maximum number of degrees of freedom in the error sum of squares.

Disadvantages.

a. The design is suited only for small numbers of treatments and for homogenous experimentai material.

b. Compared to others, this design is often less efficient for field experiments, because the experimental units for field experiments are not uniform. Therefore, the error sum of squares is inflated.

EXAMPLE: Suppose we have four treatments, T_1, T_2, T_3, and T_4, with five replications. The total number of experimental units is 20. These experimental treatments are located on the experimental site using a randomization procedure.

The data and its analysis for the four treatments are as follows.

	Treatments			
Replication	T_1	T_2	T_3	T_4
1	12	20	11	25
2	10	19	13	26

3	13	20	14	22	
4	11	19	11	27	
5	10	21	15	26	
Totals	56	99	64	126	Grand total = 345

For ease of calculating the variance, a correction factor (CF) is often used. This was particularly true before the widespread use of personal computers and statistical software programs. Examination of the formula for calculating variance

$$\frac{\Sigma (x_i - \bar{x})^2}{(n - 1)}$$

shows that with this formula each individual value must have the mean subtracted, and the differences squared and summed. Another way to accomplish the same result is to square all values and divide by the number of values (n) and then subtract this value from the sum of all individual values squared. With a hand calculator both results can be obtained easily. Since the CF is the same for all variance calculations it need be calculated only once.

1. Calculate **correction factor** (CF)

$$CF = \frac{(\Sigma x_{ij})^2}{n} = \frac{345^2}{20} = 5{,}951.24$$

x_{ij} stands for each individual treatment value (i) in the jth replication, i.e. 1, 2, 3, 4, 5.

2. Calculate **total sum of squares** (TSS)

$$\begin{aligned} TSS &= \sum x_{ij}^2 - CF \\ &= (12^2 + 10^2 + 13^2 \ldots + 26^2) - CF \\ &= 6619 - 5951.25 \\ &= 667.75 \end{aligned}$$

3. Calculate **treatment sum of squares** (Trt SS)

$$\text{Trt SS} = \frac{\Sigma x_i^2}{r} - \text{CF}$$

$$\text{Trt SS} = \frac{(56^2 + 99^2 + 64^2 + 126^2)}{5} - \text{CF}$$

$$= \frac{32{,}909}{5} - \text{CF}$$

$$= 6{,}581{,}8 - 5{,}951.25$$

$$= 630.55$$

r = number of replications

4. Calculate **error sum of squares** (ESS)

$$\text{ESS} = \text{TSS} - \text{Trt SS}$$

$$= 667.75 - 630.55$$

$$= 37.20$$

Analysis of variance table

Source	df	SS	MS	F
Treatments	(t−1) 3	630.55	210.18	90.59**
Error	t(r−1) 16	37.20	2.32	
Total	(rt−1) 19	667.75		

The mean squares (MS) are obtained by dividing the sums of squares (SS) by their respective degrees of freedom. The F-value or F-statistic is computed by dividing the treatment mean square by the error, that is: F = 210.18/2.32 = 90.59. This ratio is known as the **variance ratio**, and is denoted by F in honour of Sir Ronald A. Fisher and often called **Fisher's ratio**.

The calculated F-value is compared with the tabular F value to decide whether to accept or reject the null hypothesis of no difference between population means. Tabular F values are calculated for various levels of probability and various degrees of freedom on a theoretical basis and are used for comparison with experimental data. If calculated F values exceed tabular F values, then the results are considered to be due to some factor other than chance, in this case the treatment. Thus the values of F required for significance at different levels of probability and for different degrees of freedom for the numerator and the denominator of the ratio are obtained. Commonly, 5 and 1 percent probability levels are used for testing significance. The tabular F-values for 3 and 16 degrees of freedom in the present data are 3.24 and 5.29, respectively. Because the calculated F exceeds the

1% tabular F, we conclude that F is **highly significant**. This result is generally indicated by placing two asterisks (**) on the computed F value in the analysis of variance table.

If the computed F is larger than the tabular F value at the 0.05 probability level but smaller or equal at the 0.01 level, the treatment difference is said to be significant at the 5% level. Such a result is indicated by placing one asterisk on the computed F value. Significant or highly significant results reveal real differences among treatment means. If the computed F is less than or equal to the tabular F value at the 0.05 probability level, then the treatment difference is considered non-significant.

The standard error of a treatment mean ($s\bar{x}$) and a difference between treatment means ($s\bar{x}_i - \bar{x}_{i'}$) are useful statistics for comparing differences among treatment means and differences among pairs of treatment means. The numerical values for these are:

$$s\bar{x} = \sqrt{\frac{s^2}{r}} = \sqrt{\frac{2.32}{5}} = 0.6812$$

$$s\bar{x}_i - \bar{x}_{i'} = \sqrt{\frac{2s^2}{r}} = \sqrt{\frac{2(2.32)}{5}} = 0.9632$$

In the above formula, s^2 is the error mean square, and r is the number of replications.

Similarly, the coefficient of variability (CV) is calculated as:

$$\text{CV} = \frac{\sqrt{s^2}}{\bar{x}} \times 100 = \frac{\sqrt{2.32}}{17.25} \times 100 = 8.83\%$$

7.3.3 Randomized complete block design

This is the most widely used experimental design in agricultural experimentation. It is used when the experimental units are heterogeneous. The units are grouped in such a way that within the same group the units are homogeneous but units between groups are heterogeneous. Then treatments are assigned at random to each group of homogeneous units, and these are called **complete blocks** or **replicates**. The number of experimental units within a complete block or replicate is equal to the number of treatments being studied. The actual number of replications in the design is determined by the availability of resources and considerations of cost and precision. In general, the two main characteristic features of a randomized complete block design are:

1. the grouping of homogeneous experimental units, and
2. the random allocation of treatments separately in each block.

The randomization procedure for a randomized complete block design consists of assigning the given treatments at random to the experimental units of each replicate. A separate randomization is used for each replicate.

Advantages.

1. The statistical analysis of data is quite simple and rapid. If necessary, the data for one or more complete blocks or treatments may be omitted from the analysis without complicating it.
2. For most types of experimental work, this design yields greater accuracy than the completely randomized design.
3. Missing values are easily estimated for data analysis.
4. The design is used for a wide range of treatments, and there is no restriction on the number of replications.

Disadvantages.

1. It cannot be used for testing a large number of treatments, because an increase in the number of treatments results in an increase in heterogeneity.
2. There are fewer degrees of freedom in error than with the completely randomized design.

EXAMPLE: Suppose we have six treatments, T_1, T_2, T_3, T_4, T_5, T_6, and four replications. The total number of experimental units is 24, six for each replication. The treatments are allotted to the experimental units by the randomization procedure separately for each replication.

The data and its analysis for the six treatments are as follows:

	Treatments						
Replications	T_1	T_2	T_3	T_4	T_5	T_6	Total
1	98	89	93	75	70	77	502
2	100	94	91	80	88	81	534
3	90	98	90	69	84	91	522
4	96	91	95	78	80	89	529
Total	384	372	369	302	322	338	2087

1. Calculate **correction factor** (CF)

$$CF = \frac{(\Sigma x_{ij})^2}{t\ r} = \frac{(2087)^2}{6 \times 4} = 181{,}482.0417$$

t = number of treatments

2. Calculate **total sum of squares** (TSS)

$$\begin{aligned} TSS &= \Sigma \times {}^2_{ij} - CF \\ &= (98^2 + 89^2 + 93^2 + \ldots + 89^2) - CF \\ &= 183{,}279 - 181{,}482.0417 \\ &= 1{,}796.9583 \end{aligned}$$

3. Calculate **replication sum of squares** (RSS)

$$\begin{aligned} RSS &= \frac{\Sigma r_j^2}{t} - CF \\ &= \frac{502^2 + 534^2 + 522^2 + 529^2}{6} - CF \\ &= \frac{1{,}089{,}485}{6} - CF \\ &= 181{,}580.8333 - 181{,}482.0417 \\ &= 98.7916 \end{aligned}$$

4. Calculate **treatment sum of squares** (Trt SS)

$$\begin{aligned} TrtSS &= \frac{\Sigma t_i^2}{r} - CF \\ &= \frac{384^2 + 372^2 + 369^2 + 302^2 + 322^2 + 338^2}{4} - CF \\ &= \frac{731{,}133}{4} - CF \\ &= 182{,}783.25 - 181{,}482.0417 \\ &= 1{,}301.2083 \end{aligned}$$

5. Calculate **error sum of squares** (ESS)

$$\begin{aligned} ESS &= TSS - RSS - TrtSS \\ &= 1{,}796.9583 - 98.7916 - 1{,}301.2083 \\ &= 396.9584 \end{aligned}$$

Analysis of variance table

Source	df	SS	MS	F
Replications	(r–1) 3	98.7916	32.9305	1.24
Treatments	(t–1) 5	1,301.2083	260.2417	9.83**
Error	(r–1)(t–1) 15	396.9584	26.4639	
Total	(rt–1) 23	1,796.9583		

The F value of treatment differences with 5 and 15 degrees of freedom is highly significant. This is an indication that there are real differences among the treatment means.

The sample standard error of the difference between two equally replicated treatments ($s\bar{x}_i - \bar{x}_{i'}$) and the coefficient of variability (CV) are calculated as:

$$s\bar{x}_i - \bar{x}_{i'} = \sqrt{\frac{2s^2}{r}} = \sqrt{\frac{2(26.4639)}{4}} = 3.6376$$

$$CV = \frac{\sqrt{s^2}}{\bar{x}} \times 100 = \frac{\sqrt{26.4639}}{86.9583} \times 100 = 5.92\%$$

7.4 Heritability

Variation in any character in a segregating or mixed population is due to both genetic and environmental factors. The genetic factor is of most importance in plant breeding since it can be used to improve the population. The greater the proportion of total variability that is due to environmental factors, the more difficult it will be to select for inherited differences. If environmental variability is small in relation to genetic differences, selection will be more efficient. The inherited portion of the variability is called **heritability**. Therefore, heritability is a statistic (expressed as a percentage or proportion) that may be used to evaluate the effectiveness of selection during segregation generations following hybridization. It is a measure of the value of selection of a particular character and an index of transmissibility of the genes controlling the character.

Heritability is the ratio of genotypic variance to total (phenotypic) variance, which includes both genetic and environmental variance. At present, heritability is used in both a broad and a narrow sense. **Broad sense heritability** refers to the ratio of genetic variance to the total (phenotypic) variance within a population. The broad sense estimates of heritability are based on the estimates of genetic and phenotypic variances.

$$\text{Heritability (broad sense)} = \frac{\text{Genetic variance}}{\text{Phenotypic variance}} = \frac{V_g}{V_g + V_e} = \frac{V_g}{V_p}$$

Genetic variance can be partitioned into additive (V_A), dominant (V_D), and epistatic interaction (V_I). Of these three, the additive is by far the largest component for most quantitative characters.

Narrow sense heritability is the ratio of the additive genetic variance to the total variance within a population. Therefore:

$$\text{Heritability (narrow sense)} = \frac{\text{Additive genctic variance}}{\text{Phenotypic variance}} = \frac{V_A}{V_p}$$

The symbol for heritability in common use is h^2. It comes from the symbol h introduced in the early 1920's for the ratio of standard deviations. The square sign results from heritability involving variances. Heritability in the broad sense includes all types of genetic variance. Narrow sense heritability, on the other hand, includes only the additive type of gene action. Heritability can range from one to zero. However, actual heritabilities fall somewhere between zero and one. If the heritability is high, it means the genotype plays a more important role than the environment in determining the phenotype. Therefore, a character with a high heritability will respond to selection more than a character with low heritability. In general, plant traits controlled by a few genes have higher heritability than polygenic traits.

7.4.1 Methods of estimating heritability

Estimates of heritability of many characteristics of crop plants can be found in the literature. Although most estimates for any given character are similar, differences occur due to the particular segregating populations studied and the environment where the study is conducted. There are many ways to estimate the degree of heritability in crop plants. Important ones include parent–offspring regression, variance components from an analysis of variance, and the use of generation variances. The method using generation variances will be discussed here in detail.

In order to estimate heritability by using generation variances, the hypothetical data given in Table 7.1 will be used as an example. Two wheat cultivars, P_1 and P_2, are crossed to have six generations/populations: P_1, P_2, F_1, BC_1, BC_2, and F_2. The data on the final plant height (cm) recorded from these generations are used to calculate mean and variances which are as follows.

Table 7.1 Mean and variance of six generations/populations of wheat for plant height (cm)

Generation/ population	Number of plants	Mean	Variance
P_1	30	100	45
P_2	30	90	35
F_1 ($P_1 \times P_2$)	30	96	40
BC_1 ($F_1 \times P_1$)	40	97	100
BC_2 ($F_1 \times P_2$)	40	92	95
F_2 (F_1 ⊗)	200	95	150

a. Broad sense heritability estimates based on variances of F_1 and F_2 generations (Burton 1951).

$$\text{Broad sense heritability} = \frac{V_{F_2} - V_{F_1}}{V_{F_2}}$$

where:

V_{F_1} and V_{F_2} are the variances of F_1 and F_2 of each cross in question.

Here, the variance of F_1 is an estimate of the environmental variance (σ^2_e). Since all F_1 plants are presumed to be genetically identical, and variance of F_2 represents the genetic variance (σ^2_g) and σ^2_e, the broad sense heritability ratio equals:

$$\frac{(\sigma_g^2 + \sigma_e^2) - \sigma_e^2}{\sigma_g^2 + \sigma_e^2} = \frac{\sigma_g^2}{\sigma_g^2 + \sigma_e^2} = \frac{\text{Genetic variance}}{\text{Phenotypic variance}}$$

The variances of F_1 and F_2 are 40 and 150, respectively (Table 7.1). Therefore, broad sense heritability is:

$$\frac{150 - 40}{150} = 0.733 = 73.3\ \%$$

b. Estimation of broad sense heritability by utilizing the variability among spaced F_2 plants in relation to the variability among spaced plants of non-segregating parents (Mahmud and Kramer 1951).

$$\frac{V_{F_2} - \sqrt{V_{P_1} \times V_{P_2}}}{V_{F_2}}$$

In this method, σ_e^2 is estimated by the geometric mean of the variance of the two parental lines, i.e. variance of the first parent (V_{P_1}) and the variance of the second parent (V_{P_2}). The variances of F_2, P_1, and P_2 are 162.8040, 44.9989, and 34.1793, respectively (Table 7.1). Therefore, the estimate of broad sense heritability is:

$$\frac{150 - \sqrt{45 \times 35}}{150} = 0.735 = 73.5\%$$

c. Broad sense heritability estimates based on variances of the parental, F_1, and F_2 populations (Weber and Moorthy 1952).

$$\text{Broad sense heritability} = \frac{V_{F_2} - \sqrt[3]{(V_{P_1})\ (V_{P_2})\ (V_{F_1})}}{V_{F_2}}$$

Here, σ_e^2 is estimated by the cube root of the product of the variances for the three non-segregating generations, i.e., parent 1, parent 2, and F_1.

Using the information given in Table 7.1, the broad sense heritability is:

$$\frac{150 - \sqrt[3]{(45)(35)(40)}}{150} = 0.735 = 73.5\%$$

d. Narrow sense heritability estimates based on the variances of F_2 and the two backcross (BC_1 and BC_2) populations (Warner 1952).

$$\frac{2V_{F_2} - (V_{BC_1} + V_{BC_2})}{V_{F_2}}$$

where V_{BC_1} and V_{BC_2} are the variances of backcross 1 and backcross 2, respectively. Accordingly, the narrow sense heritability by using the information given in Table 7.1 is:

$$\frac{2\ (150) - (100 + 95)}{150} = 0.70 = 70\%$$

7.4.2 Use of heritability in plant breeding

Estimates of heritability provide information:

a. On the relative practicability of selection, i.e. high heritability in the F_2 indicates that effective selection on an individual plant basis is possible.

b. On the transmission of traits from parent to progeny. Such estimates can anticipate improvement from different types and intensities of selection.

c. To guide the plant breeder in choice of breeding systems upon estimating and comparing heritability values of different plant materials.

d. To the plant breeder faced with a problem in an unfamiliar crop or a trait about which little is known, who might find some heritability studies useful in attacking the problem more knowledgeably.

7.4.3 Genetic advance

Estimates of genetic parameters obtained from well designed and executed genetic experiments provide the breeder with the information necessary to determine the best breeding procedures for a particular species. Moreover, when heritability estimates are available, progress from selection can be predicted for any breeding system, since expected gain is a function of heritability. Therefore, such guided selection produces genetic advance. This change is of great interest to plant breeders, since it changes the population mean. The magnitude of genetic advance from selection for a character in a cross is determined by:

a. The total variation in the population from which selection is to be done, i.e., the amount of genetic variability present in the population. This variability is measured as the phenotypic variance of the population.

b. The heritability of the character concerned.

c. The selection pressure, i.e. the proportion of the population that is selected.

Genetic advance under selection (GA) is calculated by the formula:

$$GA = k \cdot \sigma_p \cdot h^2$$

where k = the selection differential in standard deviation units, σ_p = the standard deviation of the phenotypic variance of the population which is being subjected to selection, and h^2 = the estimated heritability coefficient of the trait under study.

In a selection programme, superior phenotypes are selected based upon measurement of a quantitative trait. When we select superior phenotypes, we assume that we are selecting genotypes. Therefore, progress from selection is possible only if the phenotype and genotype are correlated. The phenotypic superiority of the selected group, i.e. the difference between the mean of the selected group and the mean of the entire group, is called the **selection differential**.

The selection differential (k) is based on the mean phenotypic values of the selected lines and of the base population, the phenotypic standard deviation (σ_p), and the intensity of the selection. The intensity of selection is a statistical quantity that arises from the properties of the normal curve. This depends upon the fraction (k) of the total population and (n) of the parents selected. The selection differential is expressed in terms of standard deviation units. Therefore, the value of k varies with the intensity of selection; for example, if the highest yielding 1 percent of lines are saved, k can be shown to have the value of 2.64. Similarly, if selection intensities for plant materials are 5, 10, 15, 20, and 25 percent of the lines saved, then k takes the values of 2.063, 1.755, 1.554, 1.400, and 1.271, respectively.

If heritability is near 100 percent, little progress can be made by selection unless considerable genetic variability is present. In other words, the magnitudes of genetic advance might be equal by selecting in a cross which exhibited low heritability and high genetic variability, and in one with high heritability and low genetic variability. Therefore, estimates of heritability bear a direct relation to the response which can be expected from artificial selection.

EXAMPLE: In a sunflower population, the average head weight is 43.4 g, and the phenotypic variance and heritability are 25.93 and 0.77, respectively. If the extreme 15% of the individuals from the population are saved (k = 1.554) to raise the next generation, then the genetic advance under selection would be:

$$GA = k \cdot \sigma_p \cdot h^2$$
$$= (1.554) \cdot \sqrt{25.93} \cdot (0.77)$$
$$= 6.09 \text{ grams/plant}$$

The estimate of genetic advance has the same unit as the mean; however, it can be expressed as a percent of the population mean.

$$\text{GA (as percent of population mean)} = \frac{GA}{\bar{x}} \times 100$$

where $\bar{x}$ is the mean of the base population. In this example, the genetic advance as a percent of population mean is

$$\frac{6.09}{43.4} \times 100 = 14.03\%$$

PROBLEMS

1. What is statistics, and in what fields of study can statistics be useful?
2. Define the following sets of terms, giving a suitable example of each which illustrates the relationship between them.
   a) Population; sample
   b) Parameter; statistic
   c) Qualitative variable; quantitative variable
   d) Continuous variable; discrete variable
3. What is a sample, and how can a good sample be chosen from a population?
4. The grains per spike of seven wheat varieties were counted as 73, 80, 82, 72, 75, 66, and 85. Find the arithmetic mean of the grains per spike.
5. Find the mode and median of the data set: 19, 18, 18, and 27.
6. Define the following measures of variability by describing how each one is computed: (a) range, (b) standard deviation, (c) variance.
7. The plant height (cm) of seven wheat varieties and the number of fertile tillers are given below:

Height (cm): 75, 89, 98, 106, 65, 71, 75
Number of tillers: 14, 13, 13, 11, 8, 12, 10

Which of the two characters is more variable? Justify your answer by presenting the appropriate calculation.

8. The average seed yields of sunflower in Pakistan for eight years are: 828, 831, 830, 889, 1090, 986, 1166, and 950 kg/ha. Find the variance and the standard deviation of the seed yield.

9. Given the samples listed below, calculate for each the mean, median, mode, range, variance, standard deviation, and coefficient of variation.

   a) 11, 15, 18, 12, 8, 15
   b) 8, 19, −11, 3, 13, 2
   c) 16, 10, 12, −6, 0

10. Why is a good design important to an experiment?

11. Define and discuss each of the following terms: (a) accuracy, (b) bias, (c) precision.

12. Differentiate between the completely randomized design and the randomized complete block design.

13. What does the F-statistic measure?

14. If the F-statistic is larger than the corresponding value in the F-table, what do we conclude?

15. In a randomized complete block design with 20 wheat genotypes in three replications, the variance among the wheat genotypes for grain yield per plant (g) was 250. Complete the following ANOVA and compute a value of F to test the significance of differences among wheat genotypes. Also calculate the standard error of difference between treatment means.

Source of variation	df	Sum of squares	Mean square
Replications		360	
Genotypes			
Error			60

16. Define heritability. Give two ways to compute broad sense heritability.

17. From the information given below, calculate an estimate of broad sense heritability.
    Variance of P_1 = 240   Variance of F_1 = 360
    Variance of P_2 = 300   Variance of F_2 = 400

18. In a random-mating equilibrium population of sunflower, you estimate genetic variance for oil contents = 3.75 and broad sense heritability estimate = 0.15. If 10% of the individuals from the population are selected (k = 1.755) to intermate and form the next generation, what genetic progress would be expected?

REFERENCES

Allard, R. W. Principles of plant breeding. New York: John Wiley and Sons, Inc.; 1960.

Burton, G. W. Quantitative inheritance in pearl millet (*Pennisetum glaucum*). Agron. J. 43:409–417; 1951.

Cooper, R. A.; Weekes, A. J. Data, models and statistical analysis. New Delhi: Heritage Publishers; 1986.

Das, M. N.; Giri, N. C. Design and analysis of experiments. 2nd ed. Reprint. New Delhi: Wiley Eastern Limited; 1986.

Falconer, D. S. Introduction to quantitative genetics. 2nd ed. London: Longman Group Ltd.; 1981.

Gomez, K. A.; Gomez, A. A. Statistical procedures for agricultural research. 2nd ed. New York: John Wiley and Sons; 1984.

Hayes, H. K.; Immer, F. R.; Smith, D. C. Methods of plant breeding. 2nd ed. New York: McGraw–Hill Book Co., Inc.; 1955.

Mahmud, I.; Kramer, H. H. Segregation for yield, height and maturity following a soybean cross. Agron. J. 43:605–609; 1951.

Ostle, B.; Mensing, R. W. Statistics in research: basic concepts and techniques for research workers. 3rd ed. Ames, Iowa: The Iowa State University Press; 1975.

Panse, V. G.; Sukhatme, P. V. Statistical methods for agricultural workers. 2nd ed. New Delhi: Indian Council of Agricultural Research; 1961.

Raghavarao, D. Statistical techniques in agricultural and biological research. New Delhi: Oxford and IBH Publishing Co.; 1983.

Siegel, A. F. Statistics and data analysis: An introduction. New York: John Wiley and Sons; 1988.

Spiegel, M. R. Schaum's outline of theory and problems of statistics in SI units. Schaum's Outline Series. Singapore: McGraw–Hill International Book Company; 1981.

Steel, R. G. D.; Torrie, J. H. Principles and procedures of statistics: A biometrical approach. 2nd ed. New York: McGraw–Hill; 1980.

Warner, J. N. A method of estimating heritability. Agron. J. 44:27–430; 1952.

Weber, C. R.; Moorthy, B. R. Heritable and nonheritable relationships and variability of oil content and agronomic characters in the F_2 generation of soybean crosses. Agron. J 44:202–209; 1952.

Welsh, J. R. Fundamentals of plant genetics and breeding. New York: John Wiley and Sons; 1981.

Outline of Chapter 8

8. SEED PRODUCTION

Abdul Majid Chaudhry[1]

LEARNING OBJECTIVES

After reading this chapter, a student should be able to:

- Summarize the historical background of seed development and variety testing in Pakistan.
- Calculate the adjusted amount of seed required for a given area, based on the percentage of germination determined for a batch of seed.
- Discuss the functions of the five government departments that deal with seed production and quality control.
- Define and differentiate among prebasic seed, basic seed, certified seed, and approved seed.
- By consulting the Government minimum seed certification standards, determine the maximum percentages of off-type, weed, and other crop seeds permissible for certified seed of a given species.

The seed is the basis of plant growth. The function of seed is performed not only by the ripened ovule enclosed in the embryo of the flower. Other plant parts like the stem of ornamental plants and sugarcane, or the roots of plants such as potato also serve as the source of plant growth. These plant parts must have buds, called 'eyes', which on germination give rise to a new plant.

8.1 The background of seed production in Pakistan

Production and supply of good-quality seed is of fundamental importance in any country's agricultural development, but it is influenced by the level of production technology adopted in a particular country. In the beginning of the 20th century, farmers used their own seed and through experience

---

[1]Dr. Abdul Majid Chaudhry is former Director General, Agriculture (Research), Rice Research Institute, Kala Shah Kaku.

realized the importance of seed quality. At the time of harvesting, the farmers would save healthy and robust seeds for the next year's cultivation.

In the second phase, some attention was given to agricultural research. At this time, the focus was first on development of good-quality seed. Local material grown by the farmers was collected by research staff. The material was planted and classified into groups on the basis of plant height, maturity period, and grain quality. In what is now Pakistan, this work continued and by the end of first quarter of this century a number of pure lines were released in succession as distinct varieties of wheat, rice, cotton, and oilseed crops.

In the third phase, breeding work was started to develop better cultivars by intercrossing the pure lines already developed. In this endeavour, a number of varieties were released. Simultaneously, government farms were converted into seed farms to supplement the production efforts of the research stations. The importance of seed multiplication received real prominence when in the mid-1960's fertilizer-responsive dwarf and high-yielding varieties of wheat and rice were introduced from Mexico and the Philippines, respectively. The amazingly high yields of wheat and rice fascinated growers and the Government so much that local research received attention. Research stations were strengthened, international wheat and rice research centres providing the training and material facilities. To meet the increased demand of seed and fertilizers, the Agricultural Development and Supplies Corporation was established. Later, it was divided into two separate organizations: the Agricultural Development Supplies Corporation and the Seed Corporation. The Agricultural Development and Supplies Corporation undertook the procurement and supply of fertilizer, and the Punjab Seed Corporation undertook multiplication and distribution of seed. In other provinces also, seed corporations are being established. To regulate the multiplication of seed and release of varieties, in 1976 the Government of Pakistan enacted the Seed Act.

8.2 The Seed Act of 1976

Under the Seed Act of 1976, the following Departments were created (Irfan Ahmad 1989).

i. National Seed Council
ii. Provincial Seed Councils/Seed Corporation
iii. National Seed Registration Department
iv. Federal Seed Certification Department

I. **National Seed Council:** The National Seed Council is the chief body which supervises the operation of the Seed Act and approves the seed standards.

II. **Provincial Seed Councils:** These Councils recommend the release of crop varieties developed in their respective provinces and advise the National Seed Council on various issues relating to the improvement of the seed industry.

III. **National Seed Registration Department:** The National Seed Registration Department registers a variety on the recommendation of the Federal Seed Registration Committee (FSRC). The Committee records the morphological characters of the candidate variety which distinguish it from other existing varieties. It also ascertains the **distinctness**, **uniformity**, and **stability**, collectively termed DUS, of the variety. These parameters help in maintaining the seed quality when any variety enters into commercial production. The distinctness requirement is that a new variety must be different from already existing ones. The uniformity condition is that the seed must be sufficiently uniform to meet official standards; and the stability requirement is that the new variety must breed true from year to year.

IV. **Federal Seed Certification Department:** The Federal Seed Certification Department is responsible for controlling the quality of basic and certified seeds. It carries out the inspection of fields planted for seed multiplication and thereafter tests the lots of harvested seed. Seed lots that do not come up to the quality standards are rejected. For details, see the *Seed Certification Manual* (Irfan Ahmad 1989).

In the entire seed multiplication programme, development of new seeds is the core of the system. It is, therefore, essential to understand the system of release of a new variety.

8.3 Release of new varieties

No other technological change is as readily adopted as the seed of a new variety. Therefore, continuous breeding of varieties better than the existing ones in yield, suitability for different agro-ecological zones, and resistance to diseases and insect pests is required. A good breeding programme requires proper planning, dedicated and experienced research staff, and sufficient financial resources.

After a new strain which possesses the merits to become a future variety is developed, it is tested for two or three years at the research station. All its agronomic requirements such as seed rate, time of sowing, spacing, and fertilizer and pesticide requirements are determined. Thereafter, the breeder in collaboration with the extension staff carries out multi-locational and ecological tests.

After a target variety is found suitable, its seed is supplied to the Pakistan Agricultural Research Council (PARC) for conducting the National Uniform Trials (NUT) at different places and provinces. In addition, the National Seed Registration Department carries out DUS studies. The Provincial Seed Council approves the strain as a new variety. On clearance from the PARC Variety Approval Councils it is registered by the National Seed Council as a new variety.

8.4 Classes of seed

Under the Seed Act of 1976, the following classes of seeds are defined (Irfan Ahmad 1989):

i. Prebasic seed
ii. Basic seed
iii. Certified seed
iv. Approved seed

These categories of seed are briefly described as follows.

I. PREBASIC SEED: Prebasic seed was formally called breeder's seed. It is produced by the breeder and is of high genetic purity. Its quality is checked by the Federal Seed Certification Department in the field and in the seed lots after harvest. For identification, the bags of prebasic seed are identified with a *white tag having a diagonal violet line*.

II. BASIC SEED: Basic seed is the first progeny of prebasic seed and is multiplied on seed farms. The Federal Seed Certification Department checks the field crop to determine its suitability for further seed multiplication. Bags of basic seed are identified with *white tags*.

III. CERTIFIED SEED: Certified seed is the progeny of basic seed and is produced by registered growers selected by the Seed Corporation. The Seed Corporation, on clearance from the Federal Seed Certification Department, procures the seed and carries out the cleaning and packing at their plants. This seed is supplied to the growers.

IV. APPROVED SEED: If the supply of certified seed available is not enough to meet the demand, a second generation of multiplication is done to produce approved seed.

8.5 Seed requirement

When planning for seed production, a decision has to be made as to (a) how much seed is required and (b) how much area is to be planted under each class of seed to meet these requirements. This can be easily calculated by determining the seed rate to be used and the area to be planted. For example, if we plan to plant 100,000 hectares, and the seed rate is 100 kg/ha, we would require 10,000 tonnes of certified seed for that crop [(100,000 ha × 100 kg/ha) ÷ 1000 kg/t]. Similarly, we can work out the required quantities of basic and prebasic seed.

It is not possible to produce seed for 100 percent of the area to be sown. The percentage of the total area that is to be sown with certified seed is called the **replenishment factor**. Replenishment is planned on the basis of growers' demand and past experience of the seed used. The replenishment of wheat seed in the Punjab is 8%, whereas in Sindh it is 4%. The replenishment of paddy seed in the Punjab is 9%, while in Sindh it is 5%. For crops where the profitability is high, the replenishment is also high. The average replenishment of cotton seed for the last five years is 32% in Punjab, and about 18% in Sindh. In advanced countries like the USA, the replenishment factor ranges from 10 to 50%. The seed of cross-pollinated and often cross-pollinated crops is replenished more frequently.

For planning and producing quality seed, it is necessary to know the mode of pollination of commonly cultivated crops (Table 8.1).

Table 8.1 Mode of pollination of some important crops

Self-pollinated	Often cross-pollinated	Cross-pollinated
A. FIELD CROPS	A. FIELD CROPS	A. FIELD CROPS
Barley	Berseem	Castor
Barnyard millet	Cotton	Chicory
Black gram	Lucerne	Maize
Common millet	Red gram	Mesta
Finger millet	Sesame	Mustard
Gram	Sorghum	Niger
Green gram		Pearl millet (*bajra*)
Groundnut	B. VEGETABLE CROPS	Rapeseed
Italian millet	Brinjal	Safflower
Jute	Chili	Sunhemp
Kodo millet	Okra (*bhindi*)	Teosinte
Lathyrus (*khesari*)		
Lentil		B. VEGETABLE CROPS
Linseed		Amaranthus

Self-pollinated	Often cross-pollinated	Cross-pollinated
Little millet		Cabbage
Oat		Carrot
Paddy		Cauliflower
Soybean		Chinese cabbage
Wheat		Cucurbits
		Garden beet
B. VEGETABLE CROPS		Knol-kohl (kohlrabi)
Cluster bean		Onion
Cowpea		Spinach (*palak*)
Dolichos bean		Radish
French bean		Sugar beet
Green pea		Turnip
Kasuri *methi*		
Lettuce		
Methi		
Tomato		

Source: Nema (1985).

8.6 Seed production: precautions

For seed production the following precautions should be adopted.

i. Seed should be produced in an area representative of the crop.

ii. A variety of any crop should not be planted in a field where any other variety of the same crop was grown in the previous year. If there is a shortage of space, the same variety as planted last year should be grown. This will eliminate the danger of contamination.

iii. Fields selected for seed multiplication should be reasonably fertile and free from salts.

iv. Water supply should be assured throughout the crop season. If canal closure is expected, there must be a tubewell present as a standby water supply.

v. The crop should be planted at the best time to get optimum yields. To avoid contamination, the seed plots should be isolated from the surrounding area (see Appendix A.)

vi. A balanced fertilizer dose should be applied.

vii. The officer in charge of seed production should have a good knowledge of diseases and insect pests likely to damage the crop. If there is an attack of any pest, immediate control measures should be adopted.

viii. During the crop season, the fields should be regularly and carefully examined. Roguing of off-types and weeds should be done at three stages: vegetative, pre-flowering, and post-flowering. Seed plots should be inspected by the Federal Seed Certification Staff.

ix. The crop should be harvested at the optimum stage. Avoid early and delayed harvesting. When harvesting with combine harvesters, the seed should not be broken. Cutting the crop by hand is good for seed production.

x. After harvest, the seed should be cleaned and dried. The moisture content of grains should not exceed 13–14%. The seed should be stored in new jute bags. If old seed bags are used, they should be treated with a chemical like Malathion. The seed storage areas should be cleaned and fumigated thoroughly before storing the seed. While in storage, the seed should be fumigated if and when required. Before sowing in the next season, seeds should be treated with the recommended fungicides.

8.7 Testing seed quality

For seed quality tests, the seed should be collected in waterproof polythene bags to conserve moisture. The seed to be tested is divided into two parts—one kilogram for testing and one kilogram saved as reference sample. These samples should be labelled.

The following tests are conducted:

I. **MOISTURE CONTENT:** Moisture content is determined in a temperature-controlled oven. These days electronic moisture testers are available for quick estimation of moisture.

II. **PHYSICAL PURITY TEST:** The physical purity test involves separation of pure seed from seeds of other crops and other varieties of the same species, weed seed, and inert matter (sand, stone, mud, chaff, etc.). This is done by passing the seed through sieves. For heavy seeds, air blowers can be used. A knowledge of different varieties of the same species will be helpful in separating out the non-target seeds. A diaphanoscope, or magnifier with reflecting light, can be used to effectively screen out the seed.

III. **GERMINATION TEST:** Germination capacity is defined as "the percentage of seed capable of germinating to produce seedlings robust enough to establish in good field conditions" (Thompson 1979). Four hundred cleaned seeds are put to germination in four replications. (The minimum is 300, to be tested in three replications). Two to three layers of good quality filter paper are placed in petri dishes or in a germination box. A fine-grained sand free from gravel and toxic material can also be used. Seed-testing dishes should be thoroughly washed with hot water and soap to prevent fungal and bacterial infection. A mild solution of potassium permanganate or formaldehyde can be used for this purpose. The paper or sand should be thoroughly soaked with hot water and excess water drained off. The test seeds should be sown evenly spaced in rows. A germination test with controlled temperature is even more helpful. Normally, for summer crops a temperature of 20–25°C is required for germination. The first count of seeds germinated should be taken after four or five days, and the second after 10 days.

IV. **ADJUSTMENT OF SEED RATE:** If the germination rate is above 80%, the seed can be used. With seed having a lower germination percentage, it is advisable to adjust the quantity of seed, so that the plant population in the field does not fall below the recommended number. This adjustment of seed quantity may be calculated by the formula:

$$\text{Adjusted amount of seed} = \frac{\text{Seed rate recommended} \times 100}{\%\ \text{seed germination}}$$

EXAMPLE: If wheat seed rate is recommended @ 100 kg/ha and the percentage of seed germination is 70%, the quantity seed required will be:

$$\frac{100 \times 100}{70} = 142.85 \text{ kg}$$

Thus, instead of 100 kg, we should use 143 kg seed.

Thompson (1979) introduced the concept of analytical purity, which, when combined with germination capacity, gives a single value called **pure live value**. This value can be calculated by the formula:

$$\frac{\%\ \text{analytical purity} \times \%\ \text{germination}}{100}$$

The pure live value will indicate what percent by weight of the seed lot consists of seeds which are capable of germination. For example, if a seed lot contains 90% pure seed and the germination rate is 70%, then 63% percent of the seed will germinate, as shown in this calculation:

$$\frac{90 \times 70}{100} = 63\%$$

Now the quantity of seed required can be adjusted as explained above and again given below:

$$\frac{100 \times 100}{63} = 158 \text{ kg}$$

Instead of 100 kg, we should use 158 kg seed per hectare.

8.8 Seed dormancy

Sometimes seeds fail to germinate even when favourable moisture and temperature conditions have been provided. Such seeds are called **dormant seeds**. Copeland (1988) defined dormancy as a "physical and physiological condition of a viable seed that prevents germination even in the presence of otherwise favourable germination conditions."

The *physical* condition that prevents a dormant seed from germinating is hardness of seed coat, which makes it impermeable to water. Generally, such seeds finally germinate when the seed coat cracks or breaks in the soil under the influence of rising and falling moisture and temperature levels. Sometimes mechanical rubbing with some hard surface is required to break the seed coat. This process is called **scarification**. Soaking seeds in a low concentration of sulphuric acid helps break the seed coat. The *physiological* condition is embryo dormancy. Such seeds, particularly vegetable seeds, are pre-chilled at from 5–10°C for about a week before sowing.

There are several other causes of dormancy. An immature embryo, a result of an early harvested crop, also fails to germinate and gives the impression of embryo dormancy. Also, there are sometimes chemicals present in the outer seed layer which prevent germination. Washing with water or a weak solution of sulphuric acid drains out these injurious chemicals allowing the seeds to germinate.

8.9 Seed standards

The Federal Seed Certification Department has fixed seed standards for different crops. These standards are given in Appendices A and B.

APPENDIX A. Minimum seed certification standards (field)

Crop	Max. % undesirable seed types			% diseases by certified category		
	O.T.	O.C.	O.W.	Seed-borne	Viral	Iso. (m)
Wheat	0.10	0.05	0.02	0.20	N.A.	3
Cotton*	0.20	N.A.	N.A.	N.A.	N.A.	10
Paddy	0.10	N.A.	0.02	0.10	N.A.	3
Maize (hybrids)	0.50	N.A.	N.A.	N.A.	N.A.	200
Composite and open-pollinated varieties	1.00	N.A.	N.A.	N.A.	N.A.	200
Gram	0.50	N.A.	N.A.	1.00	N.A.	10
Rape and mustard	0.20	N.A.	0.01	0.20	N.A.	200
Cabbages (cabbage, cauliflower, chinese cabbage, knol-kohl)	0.50	N.A.	N.A.	0.50	1.00	1000
Cucurbits (ash gourd, white gourd, bitter gourd, bottle gourd, cucumber, long melon, muskmelon, pumpkin, squash gourd, red gourd, ridged gourd, sponge gourd, *tinda*, vegetable marrow, watermelon)	0.20	N.A.	none	0.20	0.50	400
Lettuce	0.20	N.A.	0.10	0.20	N.A.	25
Fenugreek	0.20	N.A.	0.10	0.20	N.A.	25
Beet root	0.20	N.A.	N.A.	N.A.	N.A.	1000
Garden pea	0.50	N.A.	N.A.	N.A.	1.00	20
Beans (cowpea, lobia, dolichos bean, French bean, cluster bean)	0.50	N.A.	N.A.	1.00	N.A.	–
Turnip, radish	0.20	N.A.	N.A.	0.20	0.20	1000
Carrot	0.20	N.A.	N.A.	N.A.	N.A.	1000
Pepper	0.30	N.A.	N.A.	0.50	0.20	200
Tomato	0.50	N.A.	N.A.	0.50	N.A.	10
Brinjal	0.20	N.A.	N.A.	0.50	N.A.	200
Okra	0.20	N.A.	N.A.	N.A.	N.A.	250
Onion	1.00	N.A.	N.A.	N.A.	N.A.	800
Coriander	0.20	N.A.	N.A.	N.A.	N.A.	500
Spinach	0.20	0.10	none	N.A.	N.A.	300
Berseem/alfalfa	1.00	N.A.	0.05	N.A.	N.A.	50
Barley/oat/triticale	0.10	0.05	0.02	N.A.	N.A.	3

Crop	Max. % undesirable seed types			% diseases by certified category		
	O.T.	O.C.	O.W.	Seed-borne	Viral	Iso. (m)
Sunflower	0.10	N.A.	N.A.	N.A.	N.A.	800
Soybean	0.50	N.A.	N.A.	N.A.	N.A.	10
Safflower	0.10	0.03	0.01	N.A.	N.A.	200
Groundnut	0.50	N.A.	N.A.	N.A.	N.A.	10
Lentil	0.10	0.01	N.A.	N.A.	N.A.	–
Jute	1.00	N.A.	N.A.	N.A.	N.A.	20
Mash/*moth*/mung	0.10	0.01	N.A.	N.A.	N.A.	3
Pearl millet	0.20	N.A.	N.A.	N.A.	N.A.	200
Sesame	0.20	N.A.	N.A.	N.A.	N.A.	25
Sorghum (open-pollinated varieties)	0.20	N.A.	none	0.10	N.A.	200
Sorghum (hybrids)	0.05	N.A.	0.05	N.A.	N.A.	200
Tobacco	0.10	N.A.	N.A.	N.A.	N.A.	400

Source: Adapted from *Seed Certification Manual* (Irfan Ahmad 1989).
O.T. = Off-types O.C. = Other-crop plants
O.W. = Objectional weed plants Iso. = Isolation distance
* – There will be no *desi* varieties in American varieties.

APPENDIX B. Minimum seed testing standards (laboratory)

Crop	P.S. min.	O.V. max.	O.C. max.	T.W. max.	O.WS. max.	I.M. max.	Germ. min.	Mst. max.	MVP. max.
Wheat	98	0.20	0.05	5/kg	N.A.	2	85	12	N.A.
Cotton	98**	0.20	N.A.	N.A.	N.A.	2	75	N.A.	N.A.
	98***		N.A.	none	N.A.	2	70	N.A.	N.A.
Paddy	98	0.10	N.A.	none	N.A.	1.50	80	12	N.A.
Maize (hybrids)	98	0.50	N.A.	none	N.A.	2	90	12	N.A.
Composite and open-pollinated varieties	98	1.00	N.A.	none	N.A.	2	90	12	N.A.
Gram	98	0.50	none	none	N.A.	2	75	9	N.A.
Rape and mustard	98	0.20	0.20	35/kg	0.01	2	80	8	5
Cabbages (cabbage, cauliflower, chinese cabbage, knol-kohl)	98	0.50	0.10	0.20	N.A.	2	70	7	5
Cucurbits (ash gourd, white gourd, bitter gourd, bottle gourd, cucumber, long melon, muskmelon, pumpkin, squash gourd, red gourd, ridged gourd, sponge gourd, *tinda*, vegetable marrow, water-melon)	98	0.20	0.10	none	none	2	60	7	6

Crop	P.S. min.	O.V. max.	O.C. max.	T.W. max.	O.WS. max.	I.M. max.	Germ. min.	Mst. max.	MVP. max.
Lettuce	98	0.20	none	0.10	N.A.	2	70	8	6
Fenugreek	98	0.20	0.10	0.10	0.10	2	65	8	7
Beet root	96	0.20	0.20	0.20	N.A.	4	65	9	8
Garden pea	98	0.50	none	none	N.A.	2	70	9	8
Beans (cowpea, lobia, doli-chos bean, French bean, cluster bean)	98	0.50	none	none	N.A.	2	75	9	8
Turnip, radish	98	0.20	0.10	0.10	N.A.	2	70	6	5
Carrot	95	0.20	0.10	0.10	N.A.	5	60	8	7
Pepper	98	0.30	0.10	0.10	N.A.	2	65	8	6
Tomato	98	0.50	none	none	N.A.	2	70	8	6
Brinjal	98	0.20	none	none	N.A.	2	70	8	6
Okra	98	0.20	none	none	N.A.	2	65	10	8
Onion	98	1.00	0.20	0.20	N.A.	2	70	8	5
Coriander	96	0.20	0.20	0.20	N.A.	4	60	9	8
Spinach	96	0.20	0.20	0.20	N.A.	3.50	65	9	8
Berseem/alfalfa	98	1.00	0.20	0.20	0.05	2	80	10	7
Barley/oat/triticale	98	0.20	0.05	0.05	N.A.	2	85	12	8
Sunflower	98	0.10	none	none	N.A.	2	80	12	8
Soybean	98	0.05	0.05	0.05	N.A.	2	80	12	8
Safflower	98	0.10	0.20	0.20	0.01	2	75	12	8
Groundnut	96	0.50	0.01	0.01	none	3	80	9	5
Lentil	98	0.10	0.01	0.10	N.A.	2	70	10	8
Jute	97	1.00	0.10	0.10	N.A.	3	75	9	N.A.
Mash/*moth*/mung	98	0.10	0.01	0.10	N.A.	2	70	10	8
Pearl millet	98	0.20	0.02	0.05	N.A.	2	70	12	8
Sesame	96	0.20	0.10	0.20	N.A.	4	75	9	8
Sorghum (open-pollinated varieties)	98	0.20	0.08	0.10	none	2	80	10	8
Sorghum (hybrids)	98	0.05	0.05	0.10	0.05	2	80	10	N.A.
Tobacco	96	0.10	0.02	0.05	N.A.	4	80	10	N.A.

O.T.	=	Off-types	O.WS.	=	Objectionable wee
O.C.	=	Other crop plants	I.M.	=	Inert matter
P.S.	=	Pure seed	Germ.	=	Germination
MVP	=	Moisture for vapour-proof containers	Mst.	=	Moisture
O.V.	=	Other distinguishable varieties	Min.	=	Minimum
O.C.	=	Other crop seeds	Max.	=	Maximum
T.W.	=	Total weeds			

** Seed standards for American cotton.

*** Seed standards for *desi* cotton.

Source: Adapted from *Seed Certification Manual* (Irfan Ahmad 1989).

QUESTIONS

1. What does *replenishment* mean in seed production?
2. Which departments deal with seed testing and registration under the Seed Act of 1976?
3. What are the main classes of seed? How is each class used?
4. Define *germination capacity*. What should normally be the minimum germination percentage?
5. What precautions should be taken in seed production?
6. Define *seed dormancy* and state some important causes of dormancy.
7. Compare the isolation distances for turnip and radish as against various beans. Why do you think there is so much difference between these two groups of crops?
8. What are the minimum acceptable germination percentages for certified wheat, cotton, and sunflower seeds?

REFERENCES

Copeland, L. O. Principles of seed science and technology. Delhi: Surjeet Publications; 1988.

Hadfield W. V. ed. Seed production, testing and distribution in European countries. Paris: Organization for European Economic Cooperation; 1954.

Hebblethwaite, P. O. ed. Seed production. Proceedings of the 28th Easter School in Agricultural Science, held in September 1978. London: Butterworth and Co.; 1980.

Irfan Ahmad, S. Seed certification manual. Islamabad: Federal Seed Certification Department, Ministry of Food and Agriculture; 1989.

Nema, N. P. Principles of seed certification and testing. New Delhi: Allied Publishers Limited; 1985.

Thompson, J. R. An introduction to seed technology. London: Leonard Hill; 1979.

Outline of Chapter 9

9. GENETIC RESOURCES AND THEIR CONSERVATION IN CROP PLANTS

A. S. Larik[1]

LEARNING OBJECTIVES

After studying this chapter, a student should be able to:

- Conduct experiments to induce genetic variability.
- Preserve seeds of different indigenous species of any crop.
- Collect seeds of different crop plants, which can be displayed at an exhibition or workshop.
- Evaluate wheat germ plasm by using different descriptors.
- Organize expeditions to explore and collect variants of species occurring in various parts of the country.
- Motivate the farmers to collect, preserve, and maintain natural variants occurring in their fields.

9.1 Introduction

Human beings have always depended on plants to meet a wide variety of needs. Such dependence has increased with advances in civilization, which has multiplied human needs. At the hunting and gathering stage of human evolution, such needs were met from certain plant species growing wild. Several species were domesticated in time and spread widely with dispersing human communities, evolving variously into distinct races under diverse environmental conditions and the different methods of cultivation adopted in various parts of the world. Crop genetic resources include such **land races**, their wild ancestors and their related species, as well as the modern advanced crop cultivars which have been created by crop breeders over the past century through the conscious application of science. The problem of erosion of crop genetic resources has arisen in part because in order to meet ever-increasing food and fibre needs aggressive agricultural extension is

---

[1]Dr. A. S. Larik is Associate Professor in the Department of Plant Breeding and Genetics, Faculty of Crop Production, Sindh Agriculture University, Tandojam.

promoting the cultivation of modern cultivars even in the remotest parts of the world, resulting in the loss of land races. The wild ancestors of crop plants and their related species are being also destroyed—through overgrazing and the cultivation of land formerly under natural vegetation. The cause for concern is the permanent loss of genetic material which could be used to introduce resistance to pests and diseases, and adverse climatic and soil conditions in modern advanced cultivars. They could also be used as a reservoir of genetic material which may be needed by future generations for breeding crops to meet new challenges posed by the highly unstable climatic conditions triggered by global warming.

9.2 Centres of origin of cultivated plants

Vavilov (1951) described the following eight independent centres of origin of cultivated plants. These centres have been elegantly discussed by Siddiqui (1959) in the context of South Asia in general, and of Pakistan in particular.

1. The Chinese centre of origin. Fruit species of *Pyrus*, *Malus prunus* and many citrus fruits have their origin in China. In all, 136 plants have been listed under this centre including among others, three species of millet (*Panicum* spp.), soybean (*Glycine max*), and many other legumes.

2. The Indian centre of origin. Although this centre excludes Pakistan, Bangladesh is included in it. This is a centre of great world importance, being the home of rice (*Oryza sativa*), sugarcane (*Saccharum officinarum*), gram (*Cicer arietinum*), pigeon pea (*Cajanus indicus*), *urd* (*Phaseolus mungo*), mung bean (*Phaseolus aureus*), cowpea (*Vigna sinensis*), hyacinth bean (*Dolichos lablab*), sword bean (*Canavalia gladiata*), *guar* (*Cyamopsis psoralioides*), *ajwain* (*Carum copticum*), brinjal (*Solanum melongena*), cucumber (*Cucumis sativus*), radish (*Raphanus sativus*), bottle gourd (*Langenaria vulgaris*), dishcloth gourd (*Luffa cylindrica*), snake gourd (*Trichosanthes anguina*), mango (*Mangifera indica*), orange (*Citrus sinensis*), tangerine, lemon (*Citrus limonia*), citron (*Citrus medica*), sour orange (*Citrus aurantium*), sour lime (*Citrus aurantifolia*), date (*Phoenix dactylifera*), correa, jackfruit (*Artocarpus heterophyllus*), wood apple (*Limonia acidissima*), *amla* (*Phyllanthus emblica*), sesame (*Sesamum indicum*), safflower (*Carthamus tinctorius*), jute (*Corchorus olitorius*), tree cotton (*Gossypium arboreum*), hemp (*Hibiscus cannabinus*), black pepper (*Piper nigrum*), betel vine, areca palm, cumin (*Cuminum odorum*), catechu (*Acacia catechu*), sandalwood (*Santalum album*), indigo (*Indigofera tinctoria*), henna (*Lawsonia alba*), senna (*Cassia senna*), *Nux vomica*, bamboo, rubber (*Hevea brasiliensis*), and others. Vavilov attributed the origin of 117 cultivated plants to this centre.

The Indo-Malayan centre of origin. This centre is yet to be thoroughly explored. It is rich in important cultivated fruits such as banana (*Musa* spp.) and pummelo (*Citrus grandis*).

3. The Central Asiatic centre of origin. This includes the whole of Pakistan, Kashmir, Afghanistan, Tajikistan, Uzbekistan, and Western Tian–shan. This centre is the birthplace of common wheat (*Triticum aestivum* L.). It is also the native home of club wheat (*T. compactum* Host), short wheat (*T. sphaerococcum* Perc.), cotton (*Gossypium herbaceum* L.), pea (*Pisum sativum*), lentil (*Lens culinaris*), bean (*Vicia* spp.), chickpea (*Cicer arietinum*), mung bean (*Vigna radiata*), mustard (*Brassica campestris*), coriander (*Coriandrum sativum*), carrot (*Daucus carota*), onion (*Allium cepa*), garlic (*Allium sativum*), spinach (*Spinacea oleracea*), portulaca (*Portulaca oleracea*), basil (*Ocimum basilicum*), pistachio (*Pistacia vera*), apricot (*Prunus armeniaca*), pear (*Pyrus pyrophorum*), almond (*Prunus amygdalus*), jujube (*ber*), grape (*Vitis vinifera*), walnut (*Juglans regia*), apple (*Pyrus malus*), and other cultivated plants of immense economic importance. Although only 47 endemic species of cultivated plants have been traced to this centre, it surpasses the Chinese centre of origin (with 136 endemic species) in the possession of plants of much higher agricultural utility.

4. The Near-Eastern centre of origin. Vavilov has attributed the origin of 81 species of cultivated plants to this centre, including nine botanical species of wheat, rye (*Secale cereale*), cherry (*Prunus cerasus*), pomegranate (*Punica granatum*), quince (*Cydonia vulgaris*), fig (*Ficus carica*), many melons (*Cucumis melo*), alfalfa (*Medicago sativa*), Persian clover (*Trifolium resupinatum*), and vetch (*Vicia sativa*).

5. The Mediterranean centre of origin. The Mediterranean area is the birthplace of some 84 cultivated plants. The notable plants of this region are olive (*Olea europaea*), the carob tree, and beet (*Beta vulgaris*).

6. The Abyssinian centre of origin. Thirty-eight endemic species of agricultural plants have been ascribed to this centre. This centre is the richest in the number of botanical varieties of wheat. Cultivated barley also owes its origin to this centre.

7. The South Mexican and Central American centre of origin. Forty-nine species have been reported to be endemic in this centre. Outstanding plants of southern Mexican and Central American origin are maize (*Zea mays*), squash (*Cucurbitae*), papaya (*Carica papaya*), cocoa (*Theobroma cacao*), and upland cotton (*Gossypium hirsutum* L.).

8. The South American (Peruvian Ecuadorean-Bolivian) centre of origin. Forty-five species of cultivated plants including dozens of important species of potato (*Solanum tuberosum* L.), tomato (*Lycopersicon esculentum*), Egyptian cotton (*Gossypium barbadense* L.), and guava (*Psidium guajava*) are natives of this centre.

The Chilean centre of origin. The common potato (*Solanum tuberosum* L.) (48 chromosomes) originated in this centre.

The Brazilian-Paraguayan centre of origin. This centre of origin has furnished to the world of agriculture such plants as manioc (*Manihot utilissima*), groundnut (*Arachis hypogaea*), rubber (*Hevea brasiliensis*), cashew (*Anacardium occidentale*), and pineapple (*Ananas comosa*).

These are the primary centres from which crop species were first moved for domestication. In view of the remarkable shifts of many crop species to distant lands for cultivation, the concept of secondary centres of diversity has emerged. These secondary centres have now become significant sources of germ plasm, especially as it has developed in relation to different biological ecosystems. In primary centres of diversity the species is in competition with other natural elements of its original environment, while in secondary centres new biological and physical stresses become selective factors. Primitive cultivars in these secondary centres are important parts of the world's germ plasm resources because of man's involvement in assuring their success.

9.3 New sources of genetic diversity

We regard the material from primary and secondary centres as the main components of germ plasm. A third and increasingly important source is germ plasm developed through highly advanced plant breeding techniques such as induced polyploidy (Khushk and Larik 1989; Siddiqui 1976), mutations (Arain and Siddiqui 1977; Larik 1978; Larik and Hafiz 1981; Siddiqui and Arain 1974; Siddiqui et al. 1979), and the combining of multiple traits into useful breeding lines.

The contribution made by plant breeders and geneticists through the manipulation of germ plasm is often overlooked in relation to the quality of our germ plasm resources. Although many of these research products would not likely survive in primary centres of germ plasm since they exert a neutral or negative influence on genetic continuity and are dependent on man's intervention to realize their benefit, they have great potential to improve yields under cultivation. The new short wheats are a perfect example of the production of a germ plasm complex that includes photoperiodic insensitivity, shortened stature through the use of the introduction 'Norin 10', broad-based disease resistance, and the ability to produce high yields with appropriate irrigation and fertilizers.

9.4 Methodology of germ plasm collection and utilization with respect to genetic resources

9.4.1 Sexually propagated species

The methodology of germ plasm collection and utilization is discussed in detail by Frankel and Brown (1984). The methodology of creation of new germ plasm is illustrated and explained in a series of publications (Larik 1978, Larik and Hafiz 1981; Larik et al. 1984a,b; 1987; Siddiqui and Arain 1974; Siddiqui et al. 1979, 1981). It has been shown that genetic variation can be induced in crops of economic importance like wheat by hybridization, mutations, or by combination of different new methods of plant breeding (Chapt. 6). The extensive experimental evidence generated by these workers has shown that in many circumstances induced variations represent a more efficient source of genetic variability than gene pools conserved by nature.

Utilization of induced variation in crop improvement is a relatively recent process, which has gained considerable momentum during the last two decades.

1. Utilization of induced variation in the production of germ plasm with improved agronomic characters. As discussed earlier, modern plant breeding is dependent on the incorporation of useful genetic variation in new cultivars. New germ plasm is constantly being generated in the Plant Genetics Division, Atomic Energy Agricultural Research Centre, Tandojam (Sindh). In wheat alone, more than 250 new genetically stable strains have been produced which are being utilized in national and international wheat breeding programmes. The team of plant geneticists headed by Dr. K.A. Siddiqui is actively engaged in developing new germ plasm in crops of economic importance such as rice (*Oryza sativa*), cotton (*Gossypium hirsutum*), soybean (*Glycine max*), mung bean (*Vigna radiata*), lentil (*Lens culinaris*), sugarcane (*Saccharum officinarum*), and banana (*Musa spp.*).

Plant breeders working in other provincial agricultural universities, agricultural research institutes, PARC, and NARC have also used germ plasm to develop improved varieties of crop plants. For all the major field crops, superior lines developed by various breeders are tested across the country in all-Pakistan coordinated trials and nurseries.

For effective use, the available germ plasm needs to be screened by a specialized team to identify donors useful in relation to specific requirements of breeders. The plant breeder should maintain a dynamic working collection, from time to time adding new genetic stocks and elite strains possessing desirable genetic variability. Examples of utilization of induced variation in the production of wheat germ plasm with improved agronomic characters are listed in Tables 9.1 and 9.2.

Table 9.1 Examples of utilization of induced variations in production of wheat (*Triticum aestivum*) germ plasm with improved agronomic characters

Improved characters	No. of mutants developed in variety			
	C–591	Nayab	Indus–66	Total
Grain yield	2	6	2	10
Early maturity	–	2	1	3
Lodging resistance	3	–	–	3
Awning	11	10	–	21
Fertilizer stress (P–N)				
a. 0–0	2	–	–	2
b. 30–60	3	–	–	3
Plant height	10	10	1	21
First internode length	12	7	1	20
Second internode length	–	4	–	4
Third internode length	–	2	4	6
First flag leaf length	3	8	2	13
First flag leaf width	2	8	–	10
Second flag leaf length	–	2	1	3
Second flag leaf width	1	2	1	4
Spike length	11	6	–	17
Spikelets/spike	3	4	–	7
Florets/spike	1	6	3	10
Grains/spike	–	13	3	16
Rust resistance	3	2	2	7
Protein content	1	1	1	3
Resistance to phytotoxicity	2	2	1	5

Source: After Siddiqui (1983).

Table 9.2 Alteration of flowering, maturity and leaf rust reaction in some contemporary wheat (*Triticum aestivum*) varieties through mutagenesis

Strain number	Origin	Treatment	Days to flower	Days to maturity	Leaf rust reaction
4	Mexi–Pak	Control	73	141	S
114	Mexi–Pak	20kR	72	137	MS
71	6134XC–271	Control	74	146	MS
105	613÷XC–271	Nf 600 Rads	82	137	MS

Strain number	Origin	Treatment	Days to flower	Days to maturity	Leaf rust reaction
120	6134XC–271	10kR	86	142	MR
73	ZA–76	Control	94	145	S
152–1	ZA–76	25kR	95	143	R
153	ZA–76	25kR	84	141	R
154	ZA–76	30kR	89	142	MR
155	ZA–76	30kR	91	143	R
158	ZA–76	Nf 750 Rads	93	141	MR
75	Nuri	Control	73	137	MR
162	Nuri	Nf 600 Rads	79	139	R
163	Nuri	Nf 750 Rads	73	138	R
76	Yecora	Control	73	142	R
164	Yecora	10kR	66	136	R
166	Yecora	20kR	71	135	R
168	Yecora	20kR	69	135	MR
169	Yecora	Nf 750 Rads	64	133	R
170	Yecora	Nf 900 Rads	64	137	MR
99	Pak–70	Control	86	139	S
156	Pak–70	20kR	84	139	MR
94	Sindh–81	Control	91	139	R
182	Sindh–81	25kR+$N_aN_3 10^{-3}$ 2 hrs	81	133	MR

S = susceptible
MS = moderately susceptible
MR = moderately resistant
R = resistant
Source: After Siddiqui (1983).

9.4.2 Magnitude of genetic variability in land races in Pakistan

1. Wheat (*Triticum aestivum* L.). The PARC–Netherlands germ plasm and the 1981 Bangor–Lyallpur University expeditions collected 1948 land races of primitive wheats from Balochistan and the mountainous and submountainous regions of Skardu (Northern Pakistan), respectively. These accessions were evaluated for selected morphogenetic characters (Rao and Javed 1987; Rao et al. 1981). All the characters showed an individualistic distributional pattern with altitude. A very wide range of variation was encountered both at accession and plant levels.

The wild diploid progenitors of wheat constitute large gene pools largely unexplored by plant breeders. Species such as *Aegilops squarrosa* (= *T. tauschii*) have contributed the D–genome which is responsible for bread-making quality. *A. squarrosa* (*T. tauschii*) also contains genes which provide resistance to leaf rust. The utilization of this germ plasm could prove beneficial if, after necessary trials, it could be reintroduced in the areas where it originates; its incorporation in breeding programmes would prove an asset to the nation.

2. Barley (*Hordeum vulgare* L.). Rao et al. (1986) evaluated 83 ancient accessions of husked barley land races from Swat and Chitral for agronomic and physiologic parameters. They observed considerable variability in plant height, flag leaf area, mature ears per plant, biomass weight, heading date, and protein content. Total protein content ranged from 6.7–12.3%. Grain yield and protein content of the accessions of various altitudes are shown in Table 9.3.

Table 9.3 Grain yield (g) and protein content (%) of primitive accessions of barley (*Hordeum vulgare* L.) of different altitudes

Range of altitude	Number of accessions	Average grain yield (g)	Average protein content (%)
Below 1800	4	45.6	6.7
1800–1900	5	25.0	7.2
1900–2000	6	20.0	7.3
2000–2100	4	16.2	9.5
2100–2200	6	30.0	9.2
2200–2300	9	18.0	9.6
2300–2400	5	24.0	10.2
2400–2500	4	18.0	10.0
2500–2600	7	25.0	10.8
2600–2700	8	20.0	11.2
2700–2800	3	45.0	11.5
2800–2900	5	32.0	10.5
2900–3000	4	35.9	11.0
*	*	*	*
*	*	*	*
*	*	*	*
3300–3400	3	15.7	12.2

Source: Rao et al. (1986:741).

Several barley accessions—C99/3 and B5A/1; B7A/1 and B9A/1—collected from the northern region of Pakistan and Nepal, respectively, contributed a high level of mildew resistance genes (Sethar and Larik 1988; Sethar et al. 1989), and can be exploited in breeding mildew resistant varieties of barley. Evaluation carried out at the Agriculture Research Institute, Tandojam (Sindh) has revealed considerable variability in barley accessions such as 'Nilum', 'Clipper', and 'Atlas–68' for growth habit, cold resistance, disease resistance, days to heading, and other yield components.

Rao et al. (1986) also studied the relationships among flowering pattern, altitude, and number of days after sowing. Figure 9.1 shows three stages of flowering in primitive barleys: 100 days (9.1a), 118 days (9.1b), and 136 days (9.1c) after sowing. Percentages of flowering and non-flowering are shown in relation to altitude ranging from 1400–3000 m at 100 m intervals. Figure 9.1a shows that by 100 days after sowing 20–40% of the barleys at high altitudes (2600–3000 m) had flowered; of the barleys at lower altitudes only 5–10% had flowered. At the higher altitudes, flowering percentage showed a linear relationship with altitude. At 118 days after sowing (Fig. 9.1b), no pronounced relationship between flower initiation and altitude was observed, and flowering was initiated at only two additional altitudes. Also, the high-altitude categories did not maintain the linear relationship with altitude that they displayed at 100 days after sowing. By 136 days after sowing (Fig. 9.1c) the flowering and non-flowering populations found at altitudes of 1700–2000 m showed reciprocal and linear relationships with altitude, respectively. Populations at altitudes of 2100–2300 m showed the same relationships, but the rates of percentage change were not constant. The non-flowering populations from 1700–3000 m have a hyperbolic relation to altitude, with a peak at 2300 m; the flowering accessions, however, do not show this pattern. At altitudes of 1600 and 2300 m, a new adaptive evolution seems to have taken place for flowering.

3. Sorghum (*Sorghum bicolor* L.). *Sorghum bicolor* L. is an ancient crop, and a highly successful immigrant in Pakistan. The crop in multiple form has contributed immensely to agriculture in Pakistan. Grain, silage, pasture, and other types of sorghum are grown. Unknown processes, including natural crossing and mutations, have resulted in an enriched germ plasm even as the crop was being grown by farmers. From Pakistan a total of 19 species of sorghum were collected by Rockefeller and the International Crops Research Institute for the Semi-arid Tropics (ICRISAT) up to 1986. At present, ICRISAT is the major repository for the world sorghum germ plasm collection with a total of 26,564 accessions (Smith and Duvick 1989).

Numerous introductions have contributed to the improvement of sorghum for both grain and forage. Within the *Sorghum* genus perhaps the most valuable germ plasm yet introduced and developed for grain purposes was contained in five accessions—'Accho Kartuho', 'Saro Kartuho', 'Depar', 'Accho Bazigar', and 'Red Janpur'—in Sindh, Pakistan. The future availability of exotic germ plasm will determine continued progress in raising the genetic yield potential of sorghum.

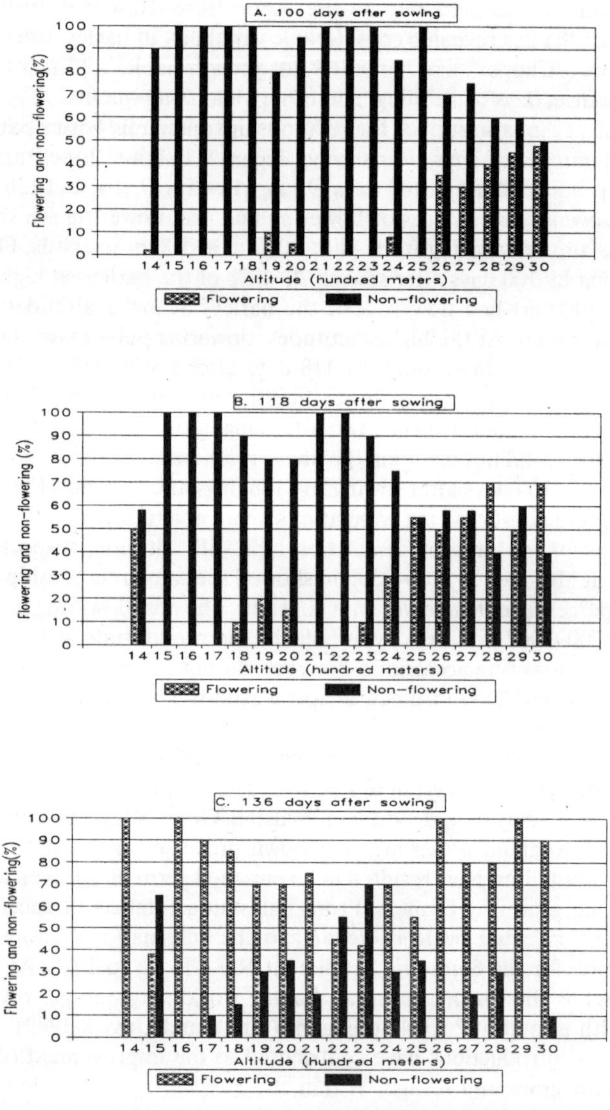

Figure 9.1 Flowering and non-flowering in husked barley (after Rao et al. 1986:739).

4. Maize (*Zea mays* L.). Maize quickly achieved world significance after the discovery of America. Presently about 13,000 separate collections of maize specimens are stored and maintained at the CIMMYT seed bank in Mexico. In Pakistan, considerable research on the improvement of maize has been carried out at various research institutes. Consequently, 13 very good synthetic/composites have been released: two in Sindh, six in Punjab, and five in NWFP.

5. Rice (*Oryza sativa*), oat (*Avena sativa*), and rye (*Secale cereale*). The germ plasm requirement for rice, oat, and rye has likewise been met to some extent by assessment and use of stocks from many countries. The exploitation of germ plasm variation for each of these crops follows a pattern similar to that discussed for wheat, barley, maize, and sorghum. The germ plasm resources for these crops and how they have been used have been elegantly reviewed by Chang et al. (1982), Fischbeck (1989), and Frankel and Bennett (1970).

9.4.3 Grasses and forage

Direct introduction continues to be fruitful for agriculture in Pakistan, but most improvements have come from the use of exotic forms in breeding programmes. Alfalfa (*Medicago* spp.) and berseem (*Trifolium* spp.) are constantly under cultivation. The *Medicago sativa* × *M. falcata* hybrids from Greece spread to Italy and from there travelled everywhere with the Roman armies. Self-pollinated varieties with high forage yield and high tolerance against inbreeding depression have been developed from crosses of *M. falcata* × *M. sativa*. Many *Medicago* species are annual. Annuals used for grazing in rotation with a cereal (wheat) crop considerably increase the returns from the land. Pakistan is one of the temperate zone countries in which several species of annual *Medicago* grow naturally. It would be timely for plant breeders to turn to study of the use of indigenous as well as introduced annual *Medicago*. Great variation in growth, vigour, maturity time, and other agriculturally significant characters can be found in both the indigenous and introduced species.

In recent years the search for forage sorghum genotypes has been intensified. The world collection at ICRISAT has been screened by several forage breeders in India, and promising lines (IS4133, IS4866, and IS11148) have been identified as suitable forage types possessing desirable attributes like plant height, profuse leafiness, high seed productivity, and comparable quality characteristics for protein and dry matter digestibility components (Prasad Rao et al. 1989). In Pakistan, in work done in Sindh, sorghum genotypes such as 'Kamandi', 'Acchi–Turi', and 'Red–Turi' have been screened and developed, so that they now possess desirable attributes for forage purposes.

9.4.4 Horticultural crops

Plant introductions have been and are used effectively in horticultural crop breeding. Scientists have turned to these germ plasm resources increasingly as specialization, mechanization, and decline of resistance to diseases and insects in commercial varieties become limiting factors for production. At the Mirpurkhas Fruit Institute in Sindh, 12 mango, 2 banana, 2 grape, 2 chiku, 2 papaya, 4 date palm, 9 citrus, and 12 vegetable cultivars have been developed, which are under intensive cultivation throughout the province.

9.4.5 New crops

New approaches and objectives in plant screening and exploration are directed at procuring germ plasm of wild species not previously considered for crop improvement. The search is for species which are promising sources of industrial end products, such as fibre and pulp from annual species, unique industrial oils, gumlike products, waxes, pharmaceuticals, insecticides, and other products with specific chemical properties. This complex effort requires the cooperation of the chemist, economic botanist, agronomist, and economist. First, the literature about a potentially useful plant group is surveyed jointly by botanists and chemists. Collections are then obtained within the genera selected for consideration.

Several potential plants such as *Carum copticum*, *Cassia fistula* L., *Cassia occidentalis* L., *Peganum harmala*, *Plantago ovata*, and *Pongamia glabra* grow wild and cultivated throughout Pakistan, and are attractive because of their unique pharmaceutical properties (Memon and Shahani 1988). Seeds and vegetative parts of these plants are used in traditional remedies for the treatment of liver and kidney trouble, chest pain, abdominal tumours, enlargement of the spleen, asthma, diarrhoea, skin diseases, chronic bronchitis, fever, and whooping cough. Agronomic characteristics of these plants such as yield and yield components need considerable improvement. They can serve as a source of a wide range of germ plasm variation and plant types which can be utilized for plant-breeding efforts directed at producing medicinally useful plants.

9.4.6 Vegetatively propagated species

Collecting vegetatively propagated plants presents the most problems. Difficulties are associated with transit survival, danger of insect and disease importation, and the high cost of propagating and developing vegetative collections. Because of these factors, many more vegetative collections than seed collections have been lost.

Among the crops most frequently collected by vegetative means are banana, pineapple, potato, sugarcane, taro, and yam. A vast array of deciduous and tropical fruit and nut crops, certain forage species, and the perennial *Arachis* species are maintained vegetatively. In addition, many ornamentals are collected as bulbs, tubers, corms, cuttings, and grafting material. In crops like potato, sugarcane, and fruits, apomixis, polyembryony, and predominant asexual reproduction are barriers to gene flow and account for the limited utilization of their germ plasm. In horticultural crops, the long generation time required to grow populations and the excessive cost of developing tree crop collections has discouraged intensive efforts. Variability is also limited for horticultural and forage crops, as there are no specific international centres and coordination.

For sweet potato (*Ipomoea batatas*), a working collection has been built at Otara Vegetable Research Station, D.S.I.R., New Zealand (John and Reitz 1971). Over a period of seven years, 175 varieties have been assembled and evaluated comparatively, and are now available to sweet potato breeders. The introductions undergo varietal identification and determination of tuber characteristics, disease screening, and quality observations. The ultimate objective is to establish a documented, disease-free working collection for use in the tropics as a source of breeding lines.

9.5 Evaluation, maintenance, and multiplication of genetic resources

9.5.1 Evaluation

Preliminary evaluation of germ plasm begins with making a sample collection of a particular crop by growing non-replicated observation rows. Complete records about the collected items are essential and are a vital part of the evaluation. Generally, deficient (e.g. diseased) plants are rejected at the collecting site. Samples of adequate size can often be evaluated without an intermediate multiplication. Single fruits, seeds, or panicles must be increased for further study, keeping in mind quarantine restrictions. Scions, cuttings, or other vegetative parts are subjected to quarantine detention and propagation. Samples should be collected from individual specimens and not be a mixture from several trees of a variety. Full evaluation is done by breeders in association with scientists from other allied disciplines and is based on both quantitative and qualitative traits. This is usually based on multi-location and multi-year trials. As information is accumulated through further evaluation and use by crop specialists, it is provided to scientists requesting material.

9.5.2 Maintenance and multiplication

A number of collections have been lost during the past few decades, thus negating the scientific and financial outlay involved in their formation. Even large centres with modern facilities and adequate expertise have reported up to 50 percent loss of samples from their original collections. Such problems can be the result of the incorrect choice of sample size for multiplication, regeneration in the wrong environments, or lack of attention to the breeding system of the crop during multiplication and regeneration. If germ plasm is maintained correctly, the genetic constitution of the population at the time of its collection can be preserved.

There are two possible approaches to the aims of (a) maximizing variability preserved for the future, and (b) making this variability available for current use.

1. Gene pools. The gene population can be maintained through mass selection, using sexual reproduction. Smith (1981) has emphasized the importance for conservation of maintaining diverse stock in each storage centre and of identifying stocks with special traits, such as dwarfing, storage protein variants, or disease resistance. Genetic variability can also be maintained through *in situ* conservation for locally adapted populations, whether of wild or weedy relatives of crop plants, or of primitive races.

2. Seed and pollen storage. It is now possible to store seeds for a very long period without significantly affecting viability by lowering temperature and moisture content. Alternatively, they can be stored by holding them fully imbibed at low temperatures; most major cereals and other crop species may be held in this way. This is the most promising method for long-term storage. At the moment, pollen storage is less successful, but when it becomes a practical proposition, space requirements will become even more modest. Techniques for storing plant parts of vegetatively propagated species are not yet very well developed.

9.5.3 Self-pollinated species

Wheat, oat, rice, barley, and other self-pollinated annual seed crops are relatively easy to handle in large numbers in a small space. Single plant hills sown with 10 seeds, or single or multiple rows spaced 30–60 cm apart and of any desired length may be used to increase the seeds of these crops. Simultaneously, many observations can be made for the purpose of description and evaluation. Winter hardiness, specific disease reaction, yield, etc. are evaluated under optimum conditions to detect varietal differences.

The working collections are kept in a room held at about 50°F and 50% relative humidity. It is necessary to re-grow accessions annually to meet the demand of breeders and others who use the seed, as our seed stocks are depleted in four to six years. **Descriptor** information is coded for each sample, including the genus, species, name or designation, accession number, cereal investigator (CI), and plant introduction (PI), source of seed, origin of variety, kernel colour, and growth habit; spike or panicle characteristics such as awn, chaff colour, pubescence, etc.; plant growth characteristics including plant height, tiller number per plant, culm diameter, and stem elongation; and yield components such as spike length, seeds per spike or panicle, spikelets per spike, and seed index. Further evaluation studies include investigation of resistance to insects, diseases, adverse environments, and production hazards.

Such information serves three functions: (1) It provides a history and basic description of each item. (2) It is a guide to authenticating items when they are re-grown to replenish the inventory. (3) It provides a basis for breeders to request certain types which are of major interest to them.

9.5.4 Cross-pollinated species

Rye (*Secale cereale*), maize (*Zea mays*), sorghum (*Sorghum bicolor*), cotton (*Gossypium hirsutum*), alfalfa (*Medicago sativa*), red clover (*Trifolium pratense*), fruit and forest trees, pasture grasses, and other crops are cross-pollinated. With sorghum, manual self-pollination can be done simply by placing a bag over the panicle at pollinating time. The bagged ears are then pollinated with bulked pollen from its own line. Some inbreeding depression will occur.

Evaluation and maintenance of long-lived species is more difficult and time consuming. Treatments to cause early flowering in perennials have made much progress in recent years thereby enabling breeders to obtain a seed generation at tolerable intervals. The perennial habit is an advantage to the plant breeder. However, one sowing or planting might persist forever, or until it is willfully destroyed. The number of items considered adequate for a collection differs from species to species. It depends upon the basic species and whether it is wild or under domestication, varying from thousands, even millions of individuals necessary to represent the crop.

In our collections, we are concerned with resistance, efficiency factors, physiological capacity, and nutritional value. Selective additions are more significant than simply numerous randomly added items, especially if gene pools from new geographical locations can be sampled. Those who propose that it is sufficient to retain only the base species would depend upon commercially available stocks, natural and induced mutations, and introgressive evolutionary forces to provide variability that might be needed later.

9.6 Need for the conservation of genetic resources in Pakistan

There is enormous need and scope for the conservation of genetic resources in Pakistan. This can be accomplished by undertaking regional, national, and international expeditions. Establishment of a national germ plasm centre is also essential for preserving the collected germ plasm and for its short-term and long-term storage. Regional centres should also be established for evaluation of the collected germ plasm. Establishment of botanical gardens can also be helpful in maintaining live collections of crop plants. The value of collections to contemporary and future workers anywhere, is enormously enhanced by documentation of the germ plasm at all stages—from site description and classification to morphological and physiological characterization, evaluation, utilization in research and breeding, and conservation.

STUDY QUESTIONS

1. Define *genetic resources*. Discuss their importance for crop improvement.

2. Discuss the importance and the scope of conservation of genetic resources in Pakistan.

3. How would you determine the location and the nature of genetic resources in the field?

4. Describe Vavilov's concept of variation.

5. Describe methods of efficient utilization of genetic resources.

6. Genetic resources activities are: (i) evaluation and collection, (ii) conservation, (iii) characterization, (iv) documentation, and (v) utilization. Discuss each of these in detail.

7. Describe the methodology of germ plasm collection and maintenance.

8. Genetic resources are classified as: (a) basic genetic resources, (b) derived genetic resources, (c) genetic resources at the molecular level. Discuss each of these global categories in detail.

9. How can the genetic resources of self-pollinated crops be maintained and multiplied?

10. Give a detailed list of descriptors used for wheat genetic resources.
11. 'Wild relatives of crops are hardy and resistant to disease, frost and drought.' Comment on this statement.
12. Cite examples of utilization of induced variation in the production of wheat germ plasm with particular reference to Pakistan.
13. What factors influence the utilization of germ plasm by plant breeders?
14. Discuss long-term programmes and facilities that should be established for the preservation of germ plasm.
15. Describe one theory of the origin of cultivated plants.
16. What are the criteria for determination of a centre of origin?
17. Discuss in detail the centres of origin of cultivated crops.
18. 'Genetic diversity is essential for crop improvement.' Comment on this statement.
19. Explain the procedures for documentation of germ plasm, and explain the importance of documentation.
20. Discuss each of the following: (a) genetic diversity, (b) genetic vulnerability, (c) gene pool, (d) pollen storage, (e) land races, (f) primitive species, (g) observable and non-observable characters, (h) gene centres.

REFERENCES

Arain, A. G.; Siddiqui, K. A. Stability parameters of wheat mutants. Environ. Exp. Bot. 17(1):13–18; 1977.

Brown, A. H. D. ed. The use of plant genetic resources. Cambridge: Cambridge University Press; 1989.

Chang, T. T.; Adair, C. R.; Johnston, T. H. The conservation and use of rice genetic resources. Advan. Agron. 35: 37–91; 1982.

Fischbeck, G. Evaluation of cereals in Europe. In: Brown, A. H. D. ed. The use of plant genetic resources. Cambridge: Cambridge University Press; 1989:p. 197–211.

Frankel, O. H.; Bennett, E. eds. Genetic resources in plants: Their exploration and conservation. International Biological Programs (IBP) Handbook No. 11. Oxford: Blackwell Scientific Publications; 1970.

Frankel, O. H.; Brown, A. H. D. Current plant genetic resources: A critical appraisal. In: Chopra, V. L.; Joshi, B. C.; Sharma, R. P.; Bansal, H. C. eds. Genetics: New frontiers. Proceedings of the 15th International Congress of Genetics, Vol. 4. Applied Genetics. New Delhi: Oxford and IHB Publishing Co.; 1984:p. 3–13.

Hawkes, J. G. The importance of wild germ plasm in plant breeding. Euphyt. 26(3):615–621; 1977.

John, L. C.; Reitz, L. P. Plant germ plasm now and for tomorrow. Advan. Agron. 23:1–50; 1971.

Khushk, A. M.; Larik, A. S. Colchiploidy in *Trifolium alexandrinum*. Pak. Phyton 1:11–16; 1989.

Larik, A. S. An evaluation of wheat mutant for useful agronomic characters. Genet. Agrar. 32(3/4):237–244; 1978.

Larik, A. S.; Hafiz, H. M. I. Wheat improvement by induced mutations. Wheat Inform. Serv. 53:35–39; 1981.

Larik, A. S.; Hafiz, H. M. I.; Kumbhar, M. B. Evaluation of wheat mutants for improved physiological efficiency. Wheat Inform. Serv. 59:41–42; 1984a.

Larik, A. S.; Hafiz, H. M. I.; Siddiqui, K. A. Grain-filling duration and grain yield relationships in wheat mutants. Nucl. 24:27–29; 1987.

Larik, A. S.; Siddiqui, K. A.; Hafiz, H. M. I.; Arain, M. H. Evaluation of wheat mutants for days to maturity. Wheat Inform. Serv. 58:22–24; 1984b.

Memon, M. I.; Shahani, N. M. Survey and domestication of wild medicinal plants of Sindh, Pakistan. Final Research Report submitted to PARC, Islamabad in 1988.

Prasad Rao, K. E.; Mengesha, M. H.; Reddy, V. G. International use of a sorghum germ plasm collection. In: Brown, A. H. D. ed. The use of plant genetic resources. Cambridge: Cambridge University Press; 1989:p.49–67

Rao, A. R.; Arshad, A. R.; Saeed, S. A. Magnitude of genetic variability in land races of husked barleys. In: Siddiqui, K. A.; Faruqui, A. M. eds. New genetical approaches to crop improvement. Tandojam: Atomic Energy Agricultural Research Centre; 1986:p. 735–743.

Rao, A. R.; Javed, M. I. Morphogenetic adaptations in some land races of Baluchistan wheats. Sarhad J. Agric. 3(4):515–523; 1987.

Rao, A. R.; Saeed, S. A.; Shamim, R. Magnitude of phenotypic variation in primitive wheat land races from Skardu region (N. Pakistan). Pak. J. Agric. Res. 28:23–30; 1981.

Sethar, H.; Larik, A. S. Genetic studies on the mildew resistance gene in a Nepali barley (*Hordeum vulgare* L.). Pak. J. Bot. 20:105–107; 1988.

Sethar, H.; Larik, A. S.; Hafiz, H. M. I. Determination of genetic control of mildew resistance in four Asian barley genotypes (*Hordeum vulgare* L.). Pak. Phyton 1:103–106; 1989.

Siddiqui, K. A. Plant introduction, its history, development and present status. Agri. Coll. Magazine 3:40–51; 1959.

Siddiqui, K. A. Improvement of bread wheat through induced mutations. Paper presented at the Wheat Production Seminar held in Lahore, Pakistan, 1976.

Siddiqui, K. A. Utilization of induced variations in wheat improvement. In: Sakamoto, S. ed. Proc. 6th International Wheat Genet. Symp., Kyoto, Japan. Kyoto: Plant Germ-Plasm Institute, Faculty of Agriculture, Kyoto University; 1983:p. 79–90.

Siddiqui, K. A.; Arain, A. G. Performance and selection for yield of wheat mutants derived from different cultivars. Euphyt. 23(3):585–590; 1974.

Siddiqui, K. A.; Arain, M. A.; Jafri, K. A. Evolution of high yielding wheat varieties through fast neutron treatment. Mutation Breeding Newsletter 18:6–7; 1981.

Siddiqui, K. A.; Rajput, M. A.; Arain, M. A.; Arain, A. G.; Jafri, K. A. Wheat improvement by induced mutations. 18th Gamma Field Symp., Ibaraki-ken, Japan. Institute of Radiation Breeding NIAS, MAFF; 1979:p. 33–47.

Smith, J. S. C.; Duvick, D. N. Germ plasm collections and the private plant breeder. In: Brown, A. H. D. ed. The use of plant genetic resources. Cambridge: Cambridge University Press; 1989: p. 17–31.

Smith, R. Principles and practice of germ plasm distribution and exchange. FAO/UNEP/IBPGR Technical Conference on Crop Genetic Resources. Rome: FAO; 1981:p. 31.

Vavilov, N. I. The origin, variation, immunity and breeding of cultivated plants. Chronica Botanica 13:1–366. New York: The Ronald Press Co.; 1951.

Outline of Chapter 10

10. IMPROVED VARIETIES OF CROP PLANTS IN PAKISTAN

Muhammad Amin Khan[1]
Iftikhar Ahmad Khan

LEARNING OBJECTIVES

After studying this chapter, the student should be able to describe the attributes of the improved varieties of wheat, rice, cotton, sugarcane, oilseeds, pulses, and fodders grown in Pakistan.

10.1 Wheat

10.1.1 Bread wheat (*Triticum aestivum*)

Wheat is a staple food grain of Pakistan grown over about 39% of its cropped area (*Economic Survey 1992–93*). It is the principal source of protein for our population and its straw is feed for livestock. The prosperity of the people of Pakistan depends to a large extent on good wheat harvests.

For these reasons, the improvement of wheat has been a major concern of the plant breeder of Pakistan. Several prominent wheat cultivars with high yields and tolerance to disease have been released from time to time since the start of agricultural research. The varieties released before the 60's were tall, did not respond adequately to the application of fertilizers, were prone to lodging, and were susceptible to rusts, smuts, or mildews. Consequently, their yields were low and wheat production in the country remained static.

In the 60's began the era of dwarf, fertilizer-responsive, disease-resistant varieties which were less prone to lodging. 'Mexi–Pak 65' was the first such variety to be released. 'Mexi–Pak 65' and similar other varieties ushered in the Green Revolution in Pakistan.

Breeding objectives for bread wheat include:

- Suitability of different maturity groups for early, mid, and late planting
- Special emphasis on high yield

---

[1]Dr. Muhammad Amin Khan and Dr. Iftikhar Ahmad Khan are Professor and Associate Professor, respectively, of the Department of Plant Breeding and Genetics, University of Agriculture, Faisalabad. The authors are grateful for the peer review by Dr. Muhammad Saleem, Associate Professor, NWFP Agricultural University, Peshawar.

- Resistance to diseases, especially rusts, smuts and bunts
- Resistance to lodging
- Drought tolerance (for rainfed areas)
- Heat tolerance
- Salt tolerance
- Wide adaptation
- Better milling, baking, and nutritional qualities

Bread wheat varieties under cultivation. The following are the main important wheat cultivars released for general cultivation and their important characteristics.

'Pak 81': Recommended for normal planting in irrigated areas. Yield 4-5 t/ha. Good baking qualities. Originally resistant to all three rusts, but lately its resistance to leaf rust seems to have broken down.

'Pirsabak 85': A sister line of 'Pak 81'. Higher-yielding than 'Pak 81', and with bolder grain. Suitable for both irrigated and rainfed areas. Yield potential about 7 t/ha. Resistant to leaf and yellow rusts. Can be planted in areas up to 2000 m above mean sea level.

'Khyber 87': Suitable for late planting, from 25 Nov. to 20 Dec., in irrigated areas. A replacement for 'Sonalika' which has now become susceptible to leaf rust. Because of its low stature it does not lodge.

'Lalazar 90': A short-duration cv. suitable for late planting. Has outyielded 'Blue Silver' and 'Faisalabad', both good early-maturing varieties.

'Pirsabak 91': An excellent cultivar for rainfed areas of NWFP, suitable for normal planting from 25 Oct. to 25 Nov. Resistant to all three rusts.

'Pasban 90': Suitable for general planting. Average yield about 4.5 t/ha. Replacement for 'Pak 81', for its better resistance to disease. Tolerant to salinity.

'Inqilab': A promising general-purpose cultivar released in 1992. Its yield is higher than 'Pak 81', which it has replaced. Optimum planting time is from 1 Nov. to 15 Dec.

'Rohtas': Approved as a general-purpose variety in 1992. Suitable for early planting in the rainfed areas. Higher yield potential than 'Lyallpur 73'. Resistant to rusts and tolerant to high temperatures at the seedling stage.

'Kaghan 1993': Normal-duration variety suitable for planting in the irrigated areas of NWFP. A replacement for 'Pirsabak 85', which has become susceptible to leaf rust. Optimum planting time late Oct. to third week of Nov. Average yield 5–6 t/ha.

10.1.2 Durum wheat (*Triticum durum*)

Durum wheat, also called macaroni wheat, mostly includes local types. It is grown over 8–10% of the wheat area, mostly in the submountainous rainfed areas. With the introduction of high-yielding, semi-dwarf, bread-wheat cultivars in the mid 60's, durum wheat has been replaced in a considerable part of its former range. Since the start of regular breeding work on durum wheat, a high-yielding variety 'Wadanak 85' was released for general cultivation. It yields slightly more than 'Pak 81'. Three more promising varieties with high yield potential and good chapati-making qualities are undergoing advanced-stage tests prior to their release for general cultivation.

10.1.3 Triticale

Triticale is a man-made cereal, the result of a cross between wheat and rye. It still has a number of problems such as dull-looking, shrunken grains, partial sterility, and low flour yield, which at present constrain its acceptance by the farmer. However, research is under way to improve its yield, spike fertility, disease resistance, grain plumpness, and fertilizer responsiveness. Success in these endeavours may enable triticale to be accepted for cultivation in the rainfed areas for its possible better disease resistance than wheat.

10.2 Rice (*Oryza sativa*)

Rice is an important food grain in Pakistan, and an important item of export. Varietal improvement in rice has played a significant role in increasing its yields. Such work is being conducted in all the provinces of Pakistan.

10.2.1 Punjab

Punjab has the largest area under rice. It is famous for its fine-quality basmati rice, grown mainly in Gujranwala, Kasur, Lahore, Sheikhupura, and Sialkot Districts. A rice research station was established at Kala Shah Kaku, near Lahore, in the mid-20's. Since 1970 it has been upgraded to the status of a research institute. The objectives of its rice breeding programme are the breeding of both fine and coarse grain varieties with the following attributes.

- High yield potential
- Early maturity
- Short stature, stiff stem, resistance to lodging
- Resistance to insect pests such as stem borer, and to diseases such as blast, bacterial blight, and stem rot
- Broad adaptation
- Good cooking qualities

Work on mutation breeding of rice was started at the Nuclear Institute for Agriculture and Biology (NIAB) at Faisalabad, after its establishment in 1969.

10.2.2 Sindh

Rice is grown in two agro-climatic zones in Sindh: Upper Sindh, and Lower Sindh. In Lower Sindh, the climate is cooler than in Upper Sindh in the rice-growing season.

Varietal improvement of rice was started in Sindh in 1922, with emphasis on pure line selection. A Rice Research Station was established at Dokri in 1938 which has since been upgraded to a Rice Research Institute. The Institute has the following objectives for breeding rice varieties.

- High yield potential
- Early maturity
- Tolerance to soil salinity and sodicity and to heat
- Better grain quality, and higher milling recovery
- Resistance to major insect pests and diseases
- Development of high-yielding, semi-dwarf, scented varieties
- Development of dual-purpose varieties for fodder and grain.

Mutation breeding of rice started in Sindh in the early 60's, with the establishment of the Atomic Energy Agricultural Research Centre at Tandojam.

In 1966 the variety IR–8–288–3–3 was introduced from the International Rice Research Institute (IRRI), Philippines, and released as 'Pak' ('IR–8') in 1969. It was a very high-yielding variety and boosted rice production considerably in a short span of four years, but it soon started losing ground because of its poor quality. However, another variety, IR–6–156, was found to be a good substitute. It was released for general cultivation in 1971 as 'Mehran 69' ('IR–6') and is still a popular variety in the coarse grain group.

10.2.3 NWFP

Rice is grown in two agro-climatic zones of NWFP: the plains of Peshawar and D.I. Khan Divisions; and the hilly areas of Malakand and Hazara Civil Divisions, Kurram Agency, and the adjacent tribal areas. In the northern mountains the climate is sub-humid monsoonal, with an annual rainfall of about 750–1000 mm. In the plains, the temperature is high during the rice-growing season.

In NWFP, the objectives for evolving rice varieties are as follows.

- High yield potential
- Short stature

- Early maturity
- Medium and long grain
- Tolerance to cold water and cool temperatures in the Swat Valley
- Tolerance to heat and drought for growing in the hot plains under water scarcity
- Tolerance to soil salinity and sodicity

Rice varietal improvement work was started at the Agricultural Research Institute, Tarnab in 1962 and shifted to D.I. Khan and Mansehra in 1964. Work was also started in the Swat Valley in 1976, with the establishment of the Agricultural Research Station, Mingora.

10.2.4 Balochistan

In Balochistan, rice is grown in Naseerabad District, which is adjacent to Larkana and Jacobabad Districts of Sindh. Thus varieties evolved for cultivation in Upper Sindh are also suitable for growing in Balochistan.

10.2.5 Rice varieties under cultivation

The following rice cultivars are under being grown in various parts of Pakistan at present:

'IR–6': An introduction from IRRI. Parentage: 'SIAM 29' × 'Dee–geo–woe–gen'. Approved in 1971. Non-aromatic. Shorter than 'IRRI–Pak'. Medium in maturity. Panicle awnless. Grain size medium and grain shape slender. Husk is speckled brown with chalky spot on margin. Good quality. Resistant to blast.

'JHONA 349': A pure-line selection approved for general cultivation in 1933. Non-aromatic. Tall, early transplanting (mid- June). Kernel long, slender, and translucent. Threshing easy, but likely to shatter if harvest delayed after ripening. Liable to lodging in heavy and fertile soils. Moderately resistant to blast.

'KS 282': Cross between 'Basmati 370' and 'IR 95'. Approved for cultivation in 1982. Non-aromatic. Medium in height. Early-maturing. Grain long and slender, kernel translucent. Resistant to blast.

'BASMATI 198': Result of hybridization between 'Basmati 370' and 'T (N) 1'. Approved in 1972 for Sahiwal District. Plants medium in size and dark-green in colour. Late in maturity. Panicle long and fully awned. Grain slender. Kernel long and aromatic. Good quality. Moderately susceptible to blast. Resistant to lodging.

'BASMATI 370': Result of pure line selection. Released for cultivation in 1933. Plants tall, late-maturing, panicle long and partially awned. Grains long, slender, and sword-shaped. Excellent cooking rice. Highly aromatic. Good quality. Easy to thresh. Shattering moderate. Susceptible to blast.

'BASMATI PAK': Cross of 'Basmati 370' and 'CM 7–6'. Approved in 1968. Plants tall, medium in maturity, panicle medium-short and partially awned. Grains slender, kernel extra-long, lightly aromatic. Quality excellent. Moderately susceptible to blast.

'KASHMIR–BAS': A derivative of 'Basmati 370'. Released for cultivation in 1977. Plants tall, flag leaf medium to long, early-maturing, panicle long and partially awned. Grains slender, aromatic. Highly susceptible to blast.

'BASMATI 385': Developed through hybridization between 'Basmati 370' and 'TN 1'. Approved for general cultivation in 1985. Plants medium-tall, early-maturing, can withstand high doses of fertilizer without lodging. Higher yielding than 'Basmati 370' and 'Basmati Pak'. Moderately resistant to stem borer. Grain slender and aromatic, with good cooking qualities.

10.3 Maize (*Zea mays L.*)

10.3.1 Maize improvement in Pakistan

Maize is mainly grown in NWFP and Punjab, and to a smaller extent in Sindh and Balochistan. It is used as food and feed, and in agro-industries. Maize improvement was started in the early 50's, based on transferring US hybrid maize seed-production technology. Several maize hybrids from the USA, Canada, and Australia were introduced in the 50's and early 60's. Some of these, such as 'US 13', 'Indiana 909', and 'US 523', were fairly high-yielding and adapted to certain maize production areas. The programme continued for several years but had to be discontinued because of limited adoption due to the following reasons.

- Our growers prefer early-maturing cultivars/hybrids which can tolerate high plant density and give fairly good grain and stover yields in spite of the use of low inputs and infrequent hoeing. The introduced hybrids were late-maturing and in several cases could not fit into the prevailing cropping patterns.
- High-quality F_1 hybrid seed production was a tedious job in those days due to lack of trained workers.
- The imported hybrids had deeply dented and mostly yellow grain, while preference was mostly for white and flint types.
- Specific hybrids are needed for specific ecological zones, and improved cultural practices are essential for high yields. When hybrids either

experienced climatic stresses, or were grown with sub-optimal cultural practices, growers were disappointed with their indifferent performance.

- Facilities for the production, distribution, and marketing of hybrid seed were not adequate to make an impact on maize production.
- Our farmers are in the habit of using seed for several generations, and therefore experienced low yields after the first generation. They were unwilling to buy fresh seed at high cost for each sowing.
- The yield superiority of hybrids over the synthetic/composite varieties of the same maturity class was not sufficiently high to justify the introduction of the hybrid maize technology.

Breeding objectives. For the reasons stated above, maize breeders are now concentrating on the development of broad-based synthetic and composite varieties. Breeding objectives for maize are high yield, early maturity, adaptation to specific ecological conditions, tolerance to soil and climatic stresses, high quality, and resistance to insect pests and diseases.

10.3.2 Local maize varieties under cultivation

The following local varieties of maize are currently under cultivation.

'SWABI WHITE': An important variety grown in Swabi, Mardan, and adjoining areas, and in parts of Attock and Hazara. Grains white, flinty, in 8–12 rows. Maturity 80–85 days. Yield potential about 3500 kg/ha under favourable conditions. 'Swabi White' is a selection from 'Silver King'.

'PESHAWAR WHITE': Grown in Peshawar District. Ears small, cylindrical. Kernels chalky in appearance, arranged in 8–10 short rows. Maturity 70–75 days. Yield potential about 2500 kg/ha. Grown under very high plant populations.

'SATHI': Generally grown in intensively cropped areas. Fits particularly well between the spring and autumn crops of potato. Plants small in size. Ears contain 6–8 rows of small, hard, deep-orange, flinty grains. Yield seldom exceeds 900 kg/ha. Highly susceptible to lodging and plant breakage.

'DOUBLE MAIZE': Generally planted in Hazara. Probably a selection from double-cross hybrids once introduced in the area. Plant type resembles 'Swabi White' except that the ears are larger and the grain dented. Matures in 115–125 days and yields about 4500 kg/ha.

'MONTGOMERY WHITE': Cultivated in parts of the Punjab, especially Sahiwal and Toba Tek Singh Districts. Plant medium-tall. Ears have 10–12 rows of white, bold, hard, flint grains. Yield about 2800–3000 kg/ha. Susceptible to *Helminthosporium*.

'Yellow Flint': A mixture of flint races. Plants small to medium-tall. Highly susceptible to leaf diseases. Ears with 6–10 rows of small, hard, yellow, flint grains. Highly susceptible to *Helminthosporium* and other stalk diseases.

10.3.3 Hybrid cultivars

Some maize hybrids were introduced from the USA in the past and extensively tested. Some hybrids have been developed through breeding programmes using exotic and local lines in various combinations. Because of the cultural practices of maize farmers during the 50's and 60's, and limitations of hybrid seed production on a large scale, maize hybrids had only limited adoption. Some of the hybrids released from time to time are described below.

'US 13': Double-cross hybrid. Grains yellow and deep-dented. Plants are medium-tall with broad, green leaves. Maturity 120–125 days. Highly susceptible to *Helminthosporium* leaf blight.

'Indiana 909': The first double-cross hybrid introduced from the corn belt of the USA. White, deep-dented grains. Maturity 120–125 days. Highly susceptible to leaf diseases.

'Double Cross No.7': Tall, with thick stem and broad, green leaves. Ears thick with 14–16 well-filled rows. Grains yellow, semi-dent. Maturity 110–125 days. Yield 4500–5000 kg/ha.

'Double Cross No.8': Derived from 75% local germ plasm and 25% temperate material. Maturity comparable to full-season, open-pollinated local varieties. Plants relatively thin, with few leaves. Long, cylindrical, bright-yellow and bold flint grains.

'Double Cross No. 59': This cultivar also contains 75% local germ plasm and 25% US corn belt, temperate material. Plants medium-tall, with short internodes and broad, dark-green leaves. Grains bold, yellow, and semi-dented. Yield about 40% higher than open-pollinated varieties. Very susceptible to *Helminthosporium* leaf blights.

'Double Cross No. 697': Plant medium-tall, with thick stems. Ears are medium-long and borne about the middle of the plants. Grains bold, light-yellow, dented. Yield potential about 7500 kg/ha. Highly susceptible to leaf and stalk diseases.

Revival of interest in maize hybrids. Breeding work is in progress to develop hybrids from local inbred lines. Short-duration hybrids, well adapted to local conditions, have been developed and tested at several locations. Such hybrids, under relatively good management, can give higher yields than open-pollinated varieties. Moreover, they have much better disease resistance, uniformity of height and maturity, and fit nicely in the prevailing cropping patterns. Two hybrids approved and released in 1993 for general cultivation in the irrigated plains of NWFP are described below.

'**QUAID**': Double-cross hybrid. Inbreds in its parentage, (AU 239 × AU 82) × (AU 71 × AU 121). Plants short and resistant to lodging. Can tolerate high plant density. Ears borne below the middle of the plant. Can be planted in spring as well as in normal season. Maturity 90–95 days. Kernel rows 14–18. Grains yellow, semi-dent. Yield potential about 7000 kg/ha. Fairly resistant to leaf diseases.

'**SHAHBAZ**': Three-way cross derived from local yellow inbred lines AU 151, AU 102, and AU 71. Plant type short, density-tolerant. Maturity 90–95 days in the plains. Can be planted in spring as well as summer. Kernel rows in the ear, 14–16. Yield potential same as 'Quaid'. Released for general cultivation in 1993.

Work in progress. Work is under way on the development of white-grained hybrids, especially double top crosses. Emphasis is on developing white-grained hybrids of short duration (90–95 days), which are resistant to leaf and stalk diseases and tolerant of common stresses.

10.3.4 Synthetics/composites

Work on the development of synthetic/composite varieties has been in progress since the 60's. Several good varieties have been developed and released for cultivation for specific production regions of the country. Some of these are described below.

'**NEELUM**': Derived from crosses between germ plasm from CIMMYT, US corn belt, and local varieties; and released for general cultivation in 1970. Tested internationally by the FAO and being cultivated in Iraq, Syria, and Jordan. Plants thick and strong, average height 2.5 m. Leaves dark-green, broad, slightly drooping at the end. Ears long, with bold, yellow, semi-dent grains arranged in 14–16 well-filled rows. Matures in about 100 days. Can be planted both in spring and summer, fairly resistant to leaf blights. Yield about 6500–7500 kg/ha.

'**AKBAR**': Dual-purpose variety, suitable for grain as well as green fodder. Height about 2.7 m. Leaves dark-green and broad. Ears are long, compact, and borne slightly above the middle of the plant. Grains are semi-dent. Resistant to *Helminthosporium*. Yield about 7500 kg/ha. Because of its wide adaptability, 'Akbar' is grown throughout Pakistan as well as over a considerable area of Iraq.

'AGAITI 72': Six-way cross between (M 14 × Pa 32), (WF 9 × W 9) and (A 495 × A 556). Short-season variety suitable for spring or late summer planting. Similar to local yellow flint in maturity but with about 40% higher yield. Has replaced 'Sathi' in the potato-growing areas. Plant height 2m. Ears medium-long, compact, with 12–14 rows of bold, yellow, dented grains; ears are borne below the middle of the plant. Yield about 4000–5500 kg/ha. Susceptible to *Helminthosporium*.

'SYNTHETIC 551': Result of selection from a cross between 'J1' and 'Agaiti 72'. Short-duration, maturing in 80–85 days. Average plant height about 2–3 m. Ears compact and longer than those of local varieties. Grains attractive, deep yellow, mostly flint.

'SADAF': White version of 'Neelum'. Ears compact, with bold, dull-white, semi-dent grains. Recommended for areas where white maize is preferred. Released for general cultivation in 1975.

'SOAN': Plant medium-tall. Ears with 12–14 well-filled rows of white, mostly flint, kernels. Matures in 80–85 days.

'PACHAITI SUFAID': A white-grain version of 'Agaiti 72' maturing in 80–85 days. Ears compact with deep-dented grains. Reasonably tolerant to soil salinity. Recommended for the hilly areas where people prefer white maize for food. Yield about 4000–5000 kg/ha.

'SYNTHETIC 66': Result of crosses between selected S_1 lines derived from 'Swabi White' and US corn belt white lines such as K64, K55, K49, and K27. Matures in 100–110 days. Plants thin, ears compact with semi-dent kernels. Yield about 5000–6000 kg/ha.

'ZIA': A medium-duration, dent-flint, white-grain variety. Developed from the cross ('Early King' × 'Payatte') × 'Changez'. Fairly well adapted to the submountainous areas of NWFP. Of same maturity as 'Swabi White', but with different plant type. Matures in 90–100 days. Yield about 5000–6000 kg/ha. Released for general cultivation in 1971 and still planted on considerable area. Since the seed has not been renewed, considerable admixture has occurred with local land races.

'CHANGEZ': Medium-tall, mid-season, white flint-dent variety developed by half-sib selection from crosses between US corn belt inbred lines such as WF9, M14, B37, WM 13 R, Hy, A 619, Oh 45, and a local variety, 'Swabi White'. Released for general cultivation in 1970 and is still planted on a considerable area in the hilly and submountainous parts of NWFP. Because of non-renewal of the seed of this variety, considerable admixture has occurred with local land races.

The new version of 'Changez' has an improved plant type, higher resistance to leaf blights, and shorter maturity (about 80 days). Its yield is about 5000–5500 kg/ha.

'KHYBER': A cross of the yellow-grained variety 'Akbar', with 'Bowman's Cole Cruk' (BCC), a white-dent variety from Kansas, USA. 'BCC' is a tall, late, high-yielding, drought-resistant variety. 'Khyber' is a tall, full-season, dent-flint cultivar recommended for early planting. 'Khyber' is very good for roasting ears. It matures in 110–120 days and yields about 7500 kg/ha It was released for general cultivation in 1974.

'SHAHEEN': Shaheen is the earliest-maturing white-grain variety. Developed from a cross between 'Zia' and the very early-maturing varieties 'Nodak' and 'Mandan' from the USA, and 'Payette' from Canada. Because of its special characteristics of cold tolerance in addition to early maturity, it is being cultivated in the high-elevation zones of NWFP and Afghanistan, where the growing seasons are short. It was released in 1974 and is recommended for cultivation at high elevation in the mountains. It is susceptible to leaf blights and rust. Its yield is about 4000 kg/ha.

'AZAM': White, semi-flint, mid-season variety, maturing in about 90 days. Developed from the cross ('Pirsabak 7930' × 'Zia') × 'Pirsabak 7930', it is of medium height with good resistance to lodging. Moderate resistance to leaf blights. Grains are bold and pearly white. Yield 5000–5500 kg/ha. Released in 1983. Considerably improved through half-sib family selection for disease resistance, uniformity, and plant type. It is a very good variety for the irrigated plains and is moderately drought tolerant.

'KISSAN 90': Another good, short to medium-duration, white-grain variety recommended for irrigated as well as good rainfed areas. Derived from a cross between a well-known experimental variety, 'Pirsabak 7930', and a leading local germ plasm, using a full-sib family selection scheme. Has short plant type, with low ear placement, and can tolerate high plant densities. Matures in about 85 days and can be grown late in some parts of NWFP, after Virginia tobacco is harvested. Moderately resistant to *Helminthosporium maydis*. Yield potential higher than that of 'Azam'. Approved for general cultivation in 1990. Also planted in the lowlands of Afghanistan.

'SARHAD YELLOW': Full-season (110–115 days), yellow-grain variety, developed from a cross between 'Vikram' × ('B57' × 'B37') × 'Akbar'. Height 2.3–2.5 m. Plant type resembles US temperate germ plasm. Ears long; kernels dent-flint, arranged in 14–18 rows on the ear. Released in 1971. An excellent yellow variety in the full-season group. Also popular in the lowland, warmer areas of Afghanistan. Yield about 7000–8000 kg/ha. 'Sarhad Yellow' is moderately resistant to leaf blights.

'SARHAD WHITE': White version of 'Sarhad Yellow'. White grains were selected from 'Sarhad Yellow' and increased and tested for yield, adaptation, and maturity. 'Sarhad White' is comparable to 'Sarhad Yellow' in yield, maturity, and adaptation, but is susceptible to leaf blights. Released for general cultivation in 1974.

'EHSAN': Mid-season, white flint variety. Developed from the cross between 'Sarhad White' and lot 81 C1MMYT material. Matures in 100 days. An outstanding variety in this maturity group. Ears of medium size, compact with 14–16 kernel rows. Even after the ears are harvested, the plant stays green and its stalks remain sweet. It has much better resistance to stalk rots and leaf blights than other varieties. Its yield is about 7000 kg/ha.

'DEHQAN': Another mid-season (95–100 days), white-grain variety approved for general cultivation in 1990. Plants thin and of medium height. Contains deep kernels on thin cobs and has therefore a high shelling coefficient. Yield range about 5000–7000 kg/ha.

'PAHARI': A short-duration variety bred specially for production in the cool mountain environment. Developed at Kaghan through half-sib family selection from the cross of 'Shaheen' × 'Pirsabak 7930', 'Pahari' has the earliness and cold tolerance of 'Shaheen' and the disease resistance of 'Pirsabak 7930'. It was released in 1992 for mid-elevation zones from 1000–1800 m above mean sea level. But it can also be recommended for late planting following Virginia tobacco. Its yield on mid elevations is about 5000–7000 kg/ha.

10.4 Sugarcane

Sugarcane is an important crop of Pakistan. It is produced in the southern, central, and northern geographical regions, which differ in climate and therefore in crop duration. The southern region comprises upper and lower Sindh. Lower Sindh has a coastal climate with moderate temperatures and long growing season, close to the tropical climatic pattern. Average cane yields and sugar recovery are therefore higher than those in the other regions. Upper Sindh has a more continental climate, with temperature extremes, and therefore lower yields. About 50% of Pakistan's sugar industry is concentrated in the southern region.

The central region is the Punjab. Sugarcane in this region is grown under a wide range of climatic and soil conditions. The region is characterized by a comparatively short growing season with hot, dry summers and cool winters.

The northern region also has a relatively short growing season, with low precipitation in summer and frost in winter. In this region, sugarcane is grown mainly in Peshawar, Mardan, Swabi, Bannu, and D. I. Khan Districts.

10.4.1 Sugarcane research establishments

The following research establishments are conducting research on sugarcane.

Sindh: Sugarcane Research Institute, Tandojam
Sugarcane Research Station, Sujawal
Sugarcane Research Station, Naudero

Punjab: Sugarcane Research Institute, Faisalabad
Sugarcane Research Sub-station, Khanpur
Sugarcane Flowering Station, Tret, Murree Hills

NWFP: Sugarcane Research Institute, Mardan
Sugarcane Sub-station, Sarai Nawrang, Bannu
Sugarcane Sub-station, D.I. Khan
Sugarcane Breeding Station, Jabban, Dargai

Sugarcane research objectives. Sugarcane research objectives include the evolving of varieties with desired economic traits and adaptation to specific ecological zones; the development of cultivation techniques for optimum sustainable returns to the grower; and the improved utilization of sugarcane and its products.

10.4.2 Sugarcane variety development in Punjab

Eight sugarcane cultivars have been selected and developed during the past four decades at the Faisalabad Sugarcane Research Institute: 'Col 29', 'Col 44', 'Col 54', 'BL 4', 'BL 19', 'L 116', 'L 118', and 'Triton'. 'Col 54' is still the major cultivar grown in the Punjab. 'BL 4' has a high sugar yield potential under favourable environment. 'L 116' is a good early-maturing cultivar, and 'Triton' is a general-purpose cultivar. 'BF 162' and 'CP 43–33' have been developed recently. 'BF 162' is early-maturing, has a good yield potential (90–160 t/ha) under moderate input conditions, and normally does not lodge. 'CP 43–33' has high sugar recovery and is quite suitable for cultivation under moisture-stress conditions.

Potential cane yields and expected sugar recovery of the cultivars mentioned above are summarized in Table 10.1.

Table 10.1 Experimental potential cane yields, expected sugar recovery percentage, and expected sugar yields of the main sugarcane cultivars in the Punjab

Name of variety	Year of release	Potential average cane yield (t/ha)	Potential sugar recovery (%)	Potential sugar yield (t/ha)
Col 29	1954	70	10.75	7.52
Col 44	1954	73	9.00	6.57
Col 54	1963	75	9.66	7.25
BL 19	1967	83	9.40	8.00
BL 4	1968	85	10.86	9.23
L 116	1973	75	11.16	8.37
L 118	1975	83	9.08	7.54
Triton	1983	85	10.25	8.71
BF 162	1990	90	11.00	9.90
CP 43–33	1990	80	11.90	9.52

Source: Malik (1990).

The actual provincial averages during 1980–89 were: average cane yield–36 t/ha; average sugar recovery–8.30%; average sugar yield–2.97 t/ha.

10.4.3 Sugarcane varieties in NWFP

The Sugar Crops Research Institute, Mardan has developed and released several varieties. Their important characteristics are given in Table 10.2.

Table 10.2 Important characteristics of sugarcane varieties cultivated in NWFP

Variety	Year of release	Cane yield (t/ha)	CCS %	Sugar yield (t/ha)	Maturity
CO 290	1933	40.00	8.50	3.40	Mid
CO 622	1953	45.00	10.50	4.72	Mid
CO 245	1957–58	50.00	10.50	5.25	Late
NCO 310	1960–61	50.55	11.00	5.56	Early
CP 48/103	1962–63	55.00	11.00	6.05	Early
CP 44/101	1974	55.00	11.33	6.23	Mid
IM 61	1981	60.00	11.34	6.80	Early
CP 51/21	1981	60.00	11.59	6.95	Early
CP 65/357	1984	60.00	13.04	7.82	Early

Variety	Year of release	Cane yield (t/ha)	CCS %	Sugar yield (t/ha)	Maturity
L 62/96	1984	65.00	12.34	8.02	Early
CMP 13	1986	70.00	12.47	8.72	Early
CO 1321	1987	68.00	12.32	8.37	Early

Source: Khan et al. (1990).

The introduction of 'CP 51–21', 'CP 65/357', and 'L 62/96' have significantly improved both cane yield and sugar recovery in the province. The latest CVS., 'CMP 13' and 'CO 1321', are becoming popular with growers for their higher yields, higher sugar recovery, and early maturity.

At the newly established Sugarcane Breeding Station, Jabban, three promising clones have been evolved recently by crossing 'CP 48/103' with 'CP 56/59'. These have been tested at several locations in Peshawar Valley for adaptation, yield, and other agronomic traits. They combine the desirable features of early maturity and high sugar yield.

10.4.4 Sugarcane varieties in Sindh

Several sugarcane cultivars were introduced in Sindh in the early 60's; 'PR 1000' became popular with growers in Lower Sindh and was rapidly adopted due to its high cane yield. 'Col 54', 'BL 4', 'L 113', 'L116', and 'L 126', introduced from Faisalabad, are also cultivated in Sindh. CVS. 'BL 4', 'L 113', and 'L 116' are known in Sindh for their early maturity and high sugar yield. 'BL 4' alone covers more than 60% of the sugarcane area in lower Sindh, while 'L 113' and 'L 116' are popular in upper Sindh. Habib Sugar Mills, Nawabshah, was able to achieve 11.25% sugar recovery due to a planned crushing programme of 'L 113', 'L 116', and 'BL 4' planted in its production zone.

Table 10.3 gives the salient attributes of the CVS. cultivated in Sindh.

Table 10.3 Salient attributes of sugarcane varieties grown in Sindh

Variety	Year of release	Cane yield (t/ha)	Sugar recovery %	Sugar yield (t/ha)
PR 1000	1965	110	10.30	11.3
Col 54	1966	80	10.24	8.2
BL 4	1974	120	11.10	13.3
L 113	1980	90	12.6	11.3
L 116	1980	95	12.6	12.0

Source: Malik (1990).

10.5 Oilseeds

Oilseed crops can be grouped into two main categories: conventional and non-conventional. Conventional oilseed crops include rapeseed and mustards (*toria*, sarson, raya, Japan rape, carinata, *taramira*), groundnut, linseed, and sesame. Non-conventional oilseed crops include sunflower, soybean, and safflower.

10.5.1 Research programme on oilseeds

The principal establishment for conducting research on oilseed crops is the Oilseed Research Institute at Faisalabad, created in December 1975. The technical programme of the Institute has the following major objectives:

- Collection, maintenance, and evaluation of the germ plasm of all oilseed crops for utilization in the breeding programme.
- Evolution of new varieties of oilseeds with high yield potential and high oil content.
- Development of new cultures of rapeseed/mustard with high seed yield and minimum content of erucic acid/linolenic acid/glucosinulates.
- Synthesis of short-statured, early-maturing, and high-yielding strains of rapeseed/mustard for *zaid kharif* cultivation.
- Evolution of local hybrids of sunflower.
- Development of composite varieties of sunflower for cultivation under a wide range of agro-climatic conditions.
- Synthesis of self-compatible strains of sunflower to eliminate dependence on insect pollinators.
- Development of short-duration, high-yielding strains of sunflower and soybean for adjustment in the cropping patterns of rice/cotton zones.
- Evaluation of soybean material for direct introduction of biotypes in various ecological zones.
- Evolution of dual-purpose strains of linseed for flax and seed.
- Use of irradiation in selection of desirable new types of sesame.
- Development of dwarf, high-yielding, annual strains of castor.
- Breeding oilseed varieties with stable and strong resistance to insect pests and diseases.

- Zonal tests for evaluating the performance of major existing varieties of oilseed crops.
- Cooperative yield trials for judging the performance of imported sunflower hybrids and promising rapeseed/mustard and soybean varieties.
- Technological evaluation of oilseed materials for selection of genotypes with desirable chemical constituents.

10.5.2 Major varieties of oilseeds

The major cultivars of oilseeds are described below.

Rape and mustard.

'YSL 1' ('Yellow Sarson L 1'): High-yielding, somewhat late-maturing. Suitable for rabi sowing under both irrigated and rainfed conditions. Seed yellow. Yields a high percentage of oil of extra purity.

'BSA' ('Brown Sarson A'): An improved strain obtained through mass selection. Slightly dwarf. Uniformly maturing. In the early stages of growth, the rosette of leaves hugs the ground. Particularly suited for cultivation under rainfed conditions as a rabi crop in Rawalpindi Division.

'DGL' ('Dark-Green Leaved'): An improved type selected out of Japan rape material and suitable for rabi/winter season sowing. Tall, with broad, dark-green leaves, late in maturity, provides late fodder and gives good seed yield as well. Average yield 1330–1460 kg/ha.

'BL 16' ('Raya L 16'): An improved early-maturing, high-yielding type, suited for *zaid kharif*/summer planting, slightly tolerant to drought/frost/pests.

'Taramira A.': A mass-selected, well-branched, uniformly-maturing, high-yielding strain suited to all dry tracts of the Punjab except Rawalpindi Division. Average yield 457–556 kg/ha.

'LGL' ('Light-Green Leaved'): Approved in 1934, this is a selection from Japan rape material and is suited for rabi/winter season. Medium-tall with broad and light-green leaves, mid-season maturity, good for fodder and seed yield. Average yield 820–940 kg/ha.

'Toria A' ('Toria Selection A'): A mass-selected cv. approved in 1936 suitable for *zaid kharif*/summer planting. Medium-tall with early, uniform maturity, and a profuse branching system. Seed is light-brown and has 2–3% higher oil content than other toria material. Average yield 980–1230 kg/ha.

'RL 18' ('Raya L 18'): Approved in 1939, suited for rabi/winter season sowing under both irrigated and barani conditions. Tall, high-yielding, late-maturing variety possessing less tolerance to frost and pests than the Napus group. Average yield 1480–1790 kg/ha.

'Poorbi Raya': Approved in 1972, suitable for *zaid kharif*/summer planting. Self-fertile. Medium in height with profuse branching. Early-maturing (90–115 days), allowing planting of late wheat. Tolerates diseases and pests; resistant to lodging. Oil content 39–42%. Average yield 980–1450 kg/ha.

'Peela Raya': Approved in 1985, suited for rabi-winter season sowing. A tall variety with profuse, alternate branching and strong fibrous stem. Seed colour golden-yellow. Possesses high resistance to insect pests and diseases with high tolerance to drought. Suitable for both irrigated and rainfed areas. Oil content 3–4% higher than 'RL 18'. Average yield around 1810–1980 kg/ha which is a 59% increase over 'RL 18'. Erucic acid content is also lower (about 34%) than 'RL 18' (about 47%).

'Raya Anmol': Approved in 1988. Dwarf and short-duration variety (85 days maturity), allowing the planting of late wheat after it. Can sustain higher temperature stress than 'Poorbi Raya'.

LL 84 (Future candidate line): A newly evolved strain of raya almost free of erucic acid. Average yield 914–1099 kg/ha.

Groundnut.

'No. 45': Approved in 1954. High-yielding, bold-seeded spreading type, suitable for cultivation in rainfed areas such as Rawalpindi Division. Mostly consumed as roasted nuts.

'No. 334': Approved in 1972. High-yielding, medium-seeded spreading type, suitable for cultivation in both rainfed and irrigated areas. A commercial type highly suitable for crushing.

'Banki': Approved in 1973. High-yielding bunch type, highly suitable for both rainfed and irrigated areas. Digging this cv. is much less expensive than the spreading types. Gives slightly less yield than 'No. 334'.

Linseed.

'T 5': Approved in 1942. White-flowered variety with dark-brown (Type 5) seeds. Fair yield. Susceptible to diseases such as wilt and rust. Average yield 1000–1250 kg/ha.

'Chandni': Approved in 1988. High-yielding variety having white flowers and light-brown seeds. Tolerant to diseases such as wilt and rust. Average yield 1300–1500 kg/ha.

Sesame.

'T 5' ('TYPE 5'): An improved type developed through selection. Medium-tall, fair-yielding CV. having bold, white seeds due to which it is preferred over black/brown-seeded types. Susceptible to wilt disease. Average yield 550–730 kg/ha.

'PB TIL 90' ('PUNJAB TIL 90'): Approved in 1990, medium-tall, high-yielding variety with white, bold seeds. Shows high tolerance (90%) to wilt disease. Average yield 590–810 kg/ha.

10.6 Cotton

Research on the improvement and development of cotton was started in the Punjab in the beginning of this century: An improved variety of upland cotton (*Gossypium hirsutum*), '4 F', was released in 1914 for general cultivation in the Punjab. It was developed from the material imported by the East India Company in 1818 and subsequently. This material, in general, was of the non-determinant type; with low ginning out-turn, short staple length, and low yield. Until 1948, cotton improvement work was confined to the utilization of the existing genetic variability. Consequently, cultivars such as '289 F', 'LSS', '289/43 F', '124 F', '199 F', and '238 F' were developed and released from time to time for commercial cultivation. These varieties gave a ginning out-turn of about 30–31%, staple length of 20–24 mm, and spinning performance from 20's to 40's (yarn count grades). With this material, our yield of seed cotton was about 442 kg/ha at the creation of Pakistan in 1947.

After Independence, in view of the great importance of cotton for the new country, a comprehensive programme of cotton improvement was started using both inter- and intra-specific crosses. As a result of this programme, variety 'AC 134' was released for general cultivation in the central districts of the Punjab in 1959. This variety was high-yielding, with better quality and high ginning out-turn. It also had the genetic background to provide tolerance against sucking insects, especially jassids. Mainly because of 'AC 134', the area under cotton in the Punjab increased tremendously to about 2.5 million ha. Several other good CVS. were also released, e.g. 'B 557', 'MNH 93', 'MS 84', 'SLH 41', 'CIM 70', 'FH 87', 'S 12', and 'Gohar 78'. Simultaneously, appropriate production technology was evolved and disseminated among the farmers. Introduction of improved varieties and appropriate production technology boosted the yields of cotton. At the same time, considerable improvement occurred in ginning out-turn and staple length. The ginning out-turn and staple length of the earlier varieties '4 F' and 'LSS' were 32% and 23 mm, respectively; the release of the current cultivars has increased the ginning out-turn to 41.3% and the staple length to 34 mm.

10.6.1 Research objectives in cotton breeding

The following objectives are kept in view while breeding cotton.

- Short duration, short stature, heat tolerance, high yield, desirable fibre characteristics to meet the needs of local industry as well as export standards.
- Resistance to insect pests and diseases.
- Tolerance to stresses such as soil salinity and sodicity, and high water table.
- Transfer from wild relatives to cultivated cultivars of such desirable traits as tolerance to insect pests, diseases, drought, cold, and soil salinity and sodicity.
- High seed-oil content.
- Heterosis, through the production of hybrids.
- Improvement of desi cotton for increased yield, ginning out-turn, and fibre quality.
- Production and multiplication of breeder seed.

Challenges in cotton breeding. Most insect pests infesting cotton in Pakistan are controlled by insecticides. The heavy use of insecticides involved in this practice is undesirable from environmental considerations. More recently, a cotton leaf virus disease is wreaking havoc with the cotton industry of Pakistan. The challenge for breeders is to develop cotton varieties resistant to insect pests and diseases.

10.6.2 Improved varieties of upland cotton (*Gossypium hirsutum*)

Important traits of improved varieties of upland cotton developed so far are described below.

'4 F': Released in 1914, this was the first upland cotton variety evolved at Faisalabad. The plant type was bushy and medium-tall, with hairy leaves. The staple length is 22.2 mm and ginning out-turn 33%. '4 F' went out of cultivation in the 1940's.

'289 F': Evolved at Faisalabad and approved for Multan District in 1928. Plant type was bushy, and staple length 25 mm.

'124 F': Direct selection from '289 F'. Slightly earlier, higher-yielding, and with longer staple length and greater ginning out-turn than '289 F'. Released for general cultivation in 1945.

'199 F': Developed from a natural hybrid in Multan in 1936. Its progeny were very variable. Out of them one hairy plant was selected, which ultimately gave rise to '199 F'. This variety is mid-season, with staple length of 25 mm and ginning out-turn of 35%.

'AC 134': Developed from the cross '148 F' × '199 F' at Faisalabad. Approved for general cultivation in 1958. High-yielding, early-maturing, jassid-resistant. Staple length 26 mm, ginning out-turn 34%.

'LASSANI II' ('L II'): Evolved by single-plant selection from an old variety. The plant is tall, profusely hairy, and resistant to jassid. Medium in maturity. Ginning out-turn 32.7%. Staple length 28 mm. Capable of spinning 65's. Approved for general cultivation in 1958.

'BS I' (13/26): A selection from the 'M 4' variety of Sindh. Fuzzy-seeded, early, drought-resistant, high-yielding. Released for general cultivation in 1962.

'MS 39': Evolved from a natural hybrid. Sympodial in habit. Early-maturing. Staple length, about 32 mm. Approved in 1970 for cultivation in Multan Zone.

'M 100': Developed from the backcross progeny of ('M 4 × Wild') × 'M 4'. Plants erect to bushy, with broad lobes and spreading-out branches. Early, staple length 26.2–27 mm. Ginning out-turn 34.5%. Approved for general cultivation in Sindh in 1963.

'QALLANDRI': Reportedly evolved from the cross ('M 4' × *G. anomalum*) × 'Karnak', and approved for cultivation in 1974. Staple length 28.6 mm. Ginning out-turn 33.5%. Micronaire value 3.5–3.7. Being grown in parts of Nawabshah, Dadu, and Hyderabad Districts of Sindh.

'SARMAST': A selection from a backcross ('M 4' × 'Acala') × 'M 4'. Released in 1975 for general cultivation in Khairpur District of Sindh. Fibre length 28.6 mm, micronaire value 3.5–3.8, ginning out-turn 35%.

'149 F': Result of hybridization between '124 F' × 'Babdel'. Released in 1971 for general cultivation in Multan Division. The plants are medium in height and susceptible to sucking insects. Early to medium in maturity and high-yielding. Staple length 27.9 mm. Ginning out-turn 34%. Fibre strength 93,000 lb/sq in. Fineness value 4.5.

'B 557': Developed through hybridization between '268 F' × 'L 5'. Released for general cultivation in 1976. Plants tall, bushy, and show resistance to the sucking insect complex. General-purpose, high-yielding variety. Lint percentage 35. Staple length 27.5 mm. Fibre strength 94,000 lb/sq in. Micronaire value 4.2.

'MS 84': Developed from a cross between 'MS 40' × 'DPL 16'. Medium in height and susceptible to sucking insects. Released in 1983. Yield comparatively low due to high-quality cotton. Staple length 33.0 mm. Lint percentage 34. Fibre strength 91,300 lb/sq in. Fineness value 3.9.

'CIM 109': Developed through hybridization between 'NIAB 78' and 'A/89 FM'. Approved for general cultivation in 1990 for Bahawalpur, Multan, and D.G. Khan Divisions. Medium-tall, high-yielding, and early-maturing. Heat-tolerant and can therefore be successfully grown in May. Staple length 27.5 mm. Lint percentage 35. Strength 92,000 lb/sq in. Fineness value 4.5.

'CIM 70': The parentage of this variety is ('149 F' × 'Coker 100 WA') × 'Coker 8314'. Approved for general cultivation in 1986. Short-statured (100–110 cm), early-maturing, high-yielding, and heat-tolerant. Because of its early maturity, wheat can easily be grown after cotton is harvested. Its lint percentage is 31.2%, staple length 29.0 mm, fineness 4.2, and strength 92,000 lb/sq in. This cultivar is highly susceptible to sucking insects.

'NIAB 86': Developed from 'Hl 9–6–2' × 'S.P. 16' and approved for general cultivation in 1990. Compared to other varieties, it has fairly good resistance to insect pets. Tall (140–160 cm), high-yielding, and early-maturing. Staple length 29 mm. Lint percentage 34.5. Strength 95,000 lb/sq in. Micronaire value 4.7.

'S 12': Result of hybridization between 'MNH 93' and 'Arizona 7203–14–4'. Approved during 1988 for general cultivation. Short-statured, early-maturing, and high-yielding. Liked by farmers as well as ginners for its high lint percentage (41.3%). Staple length is 28.2 mm and strength 94,000 lb/sq in. Micronaire value 4.2. Susceptible to cotton leaf virus.

'NIAB 78': Developed by irradiating the F_1 hybrid of 'AC 134' and 'Deltapine 16'. Approved for general cultivation in 1983. Medium tall (130–140 cm). Early-maturing and high-yielding. Can be grown on a variety of soils under a variety of climatic conditions from Faisalabad to Sindh. Lint percentage 36. Fibre length 27.5 mm. Strength 93,000 lb/sq in. Fineness 4.5.

'MNH 93': The result of a cross ('124 F' × 'Babdel') × ('MS 39' × 'Mex 12'). Released in 1980. Medium tall (100–140 cm). High-yielding. Tendency to lodge under a heavy load of bolls, due to weakness of its stem. Staple length 28.8 mm. Lint percentage 37.7. Strength 94,200 lb/sq in. Micronaire value 4.3.

'FH 87': Developed through hybridization between ('AC 134' × 'Paymaster'). Released 1988. Medium-tall (125–150 mm). Early-maturing. High-yielding. Resistant to sucking insects. Highest seed-oil percentage of all existing cotton varieties. Staple length 27.8 mm. Lint percentage 36.8. Strength 96,000 lb/sq in. Fineness value 4.2.

'SLH 41': Developed through hybridization between ('209 F' × 'Mysore American') × ('149 F × Mex 68'). Released 1984. Tall (165 cm). Resistant to some extent against sucking insects. Can successfully be grown even with low inputs. Lint percentage 36.7. Staple length 27.8 mm. Fibre strength 95,800 lb/sq in. Fineness value 4.4.

'RH I': Result of hybridization of ('LH 62' × 'W 1104'). Released 1990 in Bahawalpur Division. Medium-tall (120–150 cm). Early-maturing, allowing the sowing of wheat after its harvest in Nov.–Dec. Staple length 29.6 mm. Lint percentage 31.6. Fibre strength 103,000 lb/sq in. Micronaire value 3.9. Staple of 'RH I' is the finest and strongest of the cultivars discussed here.

'GOHAR 87': The parentage of this variety is 'B 557' × '149 F'. Approved for general cultivation in 1990. Tall (180 cm) and stout, therefore resistant to lodging. Resistant to sucking insects. Staple length 28.0 mm. Lint percentage 35. Strength 95,000 lb/sq in. Fineness value 4.3.

'GOMAL 93': Derived like most other current cotton cultivars through hybridization, between '387 F' and 'A 134', and selection of desirable plants from the progeny. Approved for general cultivation in D.I. Khan District of NWFP in 1994. Plant type semi-bushy, with sympodial growth habit. Lint percentage 34. Staple length 27.5 mm. Fineness value 4.5.

10.6.3 Improved varieties of desi cotton (*Gossypium arboreum* L.)

About 10% of the cotton area of Pakistan is under desi types of cotton. They have shorter staple length, yield coarser thread, and have relatively higher ginning out-turn than upland cotton varieties. Some improved varieties of desi cotton have been developed through breeding for specific purposes. A few of the important ones are described below.

'39 MOLLISONI': Released in 1934, because of its higher yield compared to an earlier variety, '15 Mollisoni'. Its staple length is 17.8 mm, and its ginning out-turn 36%.

'119 SANGUINEUN': Approved in 1941 for general cultivation in Multan to replace the local variety, which was a mixture of many biotypes. '119 S' had a higher yield and matured earlier. Staple length 17.5 mm. Ginning out-turn 35%.

'231 R': Originally developed in India. Released in the Punjab in 1959 because of its higher yield and superior fibre qualities compared to '39 Mollisoni'. Ginning out-turn 36%, staple length 12.7–15.9 mm, micronaire value 7–9.

'D 9': A reselection from 'Haroonabadi Desi'. Released in 1971. A sympodial, early-maturing variety with broad leaves and white flowers. The fibre is very short (9.5–15.9 mm). Ginning out-turn 40%. A rough and harsh variety recommended for the submountainous districts of the Punjab.

'TD 1': Also a sympodial type with broad leaves and white flowers. Fibre even harsher than that of 'D 9'. Fibre length 9.5–15.9 mm. Grown in parts of Khairpur and Nawabshah Districts of Sindh.

'S.K.D. 10/19': A reselection from 'TD 1'. Fibre slightly harsher than that of 'TD 1'. Yield higher than 'TD 1'. Ginning out-turn 40%. Released in 1975 for cultivation in Sindh.

'Ravi': A selection from 'D 465'. Approved for cultivation in 1982. Early-maturing, enabling the sowing of wheat after its harvest. Yield higher than 'D 9'. Comparatively resistant to insect pests and diseases. Ginning out-turn 41%. Staple length 14.5 mm. Micronaire value 8.1. Recommended for rainfed as well as basic cotton zone of the Punjab.

'Rohi': Developed by crossing Haroonabad local and 'D 9'. Released in 1986. Recommended for parts of Bahawalnagar District. Leaves hairy, discouraging the attack of sucking insects. High-yielding. Ginning out-turn 40%. Staple length 14 mm. Micronaire value 7.9.

10.7 Pulses

Pulses belong to the family Leguminosae. They are cultivated for their edible seeds. The more important pulses grown in Pakistan are gram or chickpea (*Cicer arietinum*), pigeonpea (*Cajanus cajan*), black gram or *mash* (*Vigna mungo*), green gram or mung (*V. radiata*), lentil (*Lens culinaris*), and horse gram (*Dolichos biflorus*). Their seeds have a high percentage of protein and form a significant part of the diet of the people of South and Southeast Asia. Some pulses, e.g. gram, black gram, and horse gram are also used as livestock feed. Gram or chickpea is grown on a larger area than any other pulse crop in Pakistan. Only recently in Pakistan has attention been given to the improvement of soybean, which is rich in protein as well as oil.

The breeding methods used for the development and improvement of pulse crops are those used for self-pollinated crops. Breeders of pulse crops in Pakistan have used one or more of the following methods for breeding varieties suitable for cultivation in their respective regions of production: introduction, selection, and hybridization; and irradiation and polyploidy where it is necessary to increase genetic variability.

10.7.1 Research objectives

The major research objectives for breeding pulse crops are as follows.

- Development of varieties of gram, lentil, mung, and *mash* with high yield potential, responsiveness to inputs, and early maturity; resistance to *Ascochyta* blight and wilt in gram; rusts, blight, and *Botrytis* in lentil; and yellow mosaic and leaf crinkle in mung and *mash*.
- Determination of economic inputs for optimum yields.
- Development of economic measures for controlling insect pests and diseases.
- Screening of *Rhizobium* strains to increase efficiency in nitrogen fixation.
- Production of prebasic and basic seed.
- Promotion of the production of pulses.

10.7.2 Varieties of pulses

The following varieties of pulses have been developed and released for cultivation.

Chickpea (***Cicer arietinum*** **L.**).

'Punjab 1': A kabuli variety developed for general cultivation in 1926. Also suitable for irrigated tracts.

'Punjab 7': A desi variety released in 1926. Resistant to wilt.

'F 8': A desi variety released in 1940. Possesses resistance against *Ascochyta* blight.

'C 12/34': A desi variety possessing resistance to both wilt and blight. Released for cultivation in 1942.

'C 612': A desi variety suitable for planting in irrigated areas. Released in 1952.

'C 727': A desi variety suitable for planting in both rainfed and irrigated areas. Released in 1964.

'C 44': A bold-seeded desi variety with high yield potential. Resistant to *Ascochyta* blight. Approved in 1983. Widely adopted for cultivation in Thal.

'Punjab 91': A desi variety selected from the segregating generations of a hybrid between 'RC 32' and 'NEC 138–2'. Approved for general cultivation in 1991. Bushy plant type. Bold-seeded. Tolerant to *Ascochyta* blight. Yield potential about 3 t/ha.

'PAIDAR 91': A selection from the cross 'C 235' × 'ILC 191'. Released in 1991. Plant type erect. Strongly resistant to *Ascochyta* blight and recommended for areas prone to frequent blight epidemics. Very good for fertile soils and under high input conditions.

'NOOR 91': A kabuli type. Developed as a result of selection from the variety 'Elip 81–293 C', which, in turn, was selected from a cross between 'ILC 191' and 'ILC 495'. Plant type semi-erect. Bold-seeded. Suitable for irrigated and high rainfall rainfed areas. Tolerant to *Ascochyta* blight. Yield potential, about 2.5 t/ha.

'NIFA 88' AND 'KARAK 1': Recently released for general cultivation in NWFP.

Lentil (*Lens culinaris* Medik).

'MASOOR 85': Released in 1985. Popular variety grown all over the lentil belt from Sialkot to Rawalpindi. High yield potential. Wide adaptation. Resistant to root rot and *Botrytis*.

Mash (*Vigna mungo* (L.) Hepper).

'MASH 48': A non-hairy type. Released in 1946 for planting, preferably in the irrigated areas.

'MASH 80': A hairy variety. High yield potential. Suitable for planting in rainfed areas. Approved in 1946.

'MASH 88': A hairy variety. High yield potential. Resistant to diseases. Released in 1990 for commercial production. Very popular in Sialkot.

Mung (*Vigna radiata* (L.) Wilczek).

'TYPE 54': A tall and long-duration variety. Approved for cultivation in 1943.

'TYPE 305': Another tall and late-maturing variety. Released in 1943.

'6601': A bushy, early-maturing, and high-yielding variety. Approved in 1971.

'NIAB MUNG 28': A mutant of local strain 'PK 17'. Approved for general cultivation in 1983. Medium-statured. Uniform maturity in 75–80 days. Matures 10 days earlier than the parent and yields 15% higher. Moderately tolerant to yellow mosaic virus. Yield about 1050 kg/ha. Widely adapted to cultivation in spring and summer.

'NIAB MUNG 121 25': A mutant of the local strain 'RC 71–27'. Approved for general cultivation in 1986. Matures three weeks earlier and yields 34% more than the parent. Yields about 1400 kg/ha and matures in about 65–70 days. Wide adaptability for planting in spring and summer. Great tolerance to yellow mosaic virus. Amenable to mechanical harvesting.

'NIAB MUNG 19–91': A mutant of the local strain 'PK 22'. Approved for general cultivation in 1986. Short-statured and top-bunching. Early and uniform maturity. Matures four weeks earlier and yields 30% higher than the parent. Matures in about 60–65 days. Yield about 1300 kg/ha. Widely adapted. Resistant to yellow mosaic virus. Suitable for intercropping, rotation with cotton, and mechanical harvesting.

'NIAB MUNG 20–21': A mutant of the local strain 'PK 22'. Short-statured, top-bunching, and resistant to lodging. Matures four weeks earlier than the parent and yields 26% higher. Wide adaptability to various growing seasons. Resistant to yellow mosaic virus. Matures in 55–60 days and yields about 1250 kg/ha. Suitable for intercropping in orchards and as a catch crop in the fallow between wheat harvest and rice/maize planting. Amenable to mechanical harvesting.

'NIAB MUNG 13–1': Developed through irradiating the seed of a cross between the local variety '6601', and an exotic variety 'VC 1972'. Bold-seeded. Top-bunching. Profusely hairy. Stiff stem. Non-lodging. Uniform maturity. Non-shattering pods. Resistant to yellow mosaic virus. Tolerant to *Cercospora* leaf spot disease. Drought-tolerant. Amenable to mechanical harvesting. Widely adapted to spring and summer planting. Maturity 65–70 days. Yield about 1550 kg/ha, over 2000 kg/ha in Thal.

'NIAB MUNG 54': Also developed through irradiation of the seed from the cross mentioned above. Extra-large seed size. Uniform maturity. Pods do not shatter. Tolerant to yellow mosaic virus and *Cercospora* leaf spot diseases. Wide adaptability to spring and summer crop seasons. Suitable for intercropping and mechanical harvesting. Maturity 70–75 days. Yield is about 1500 kg/ha.

10.7.3 Future research

The following research programme is proposed for the next 10 years.

Chickpea (*Cicer arietinum* L.)

- Screening of sources of resistance to various races of *Ascochyta rabiei* for the development of varieties resistant to blight.
- Development of varieties for: resistance to wilt/root rot; tolerance to drought; wide adaptability; erect habit and resistance to lodging under

high inputs; resistance to iron deficiency chlorosis under high inputs; early maturity; high efficiency in N fixation; high nutritional quality.

- Development of varieties with a combination of yield components through component breeding.

Lentil (*Lens culinaris* Medik)

- Development of varieties with: components of high yield potential through component breeding; wide adaptability; resistance to blight, rust, wilt/root rot, and stem rot; early maturity; less liability to shattering; resistance to lodging; high efficiency in N fixation; better cooking qualities.

Mung (*Vigna radiata* (L.) Wilczek) and *mash* (*Vigna mungo* (L.) Hepper)

- Development of varieties with: high yield potential; short duration; indeterminate habit; uniform maturity; resistance to diseases; less liability to shattering; high percentage of protein; better cooking qualities.

10.8 Fodders

Fodder crops are grown throughout the country year round mostly under irrigated conditions. About 15% of the cropped area is occupied by winter (rabi) and summer (kharif) fodders. The crop is cut by hand and carried and fed to livestock fresh. Manual labour is engaged for this purpose throughout the growing season.

10.8.1 Summer (kharif) fodders

The major summer fodders are maize, sorghum, pearl millet, and sadabahar (sorghum × Sudan grass). They may be grown pure or mixed with legumes such as cowpea or *guara*. They may also be grown in combinations of two compatible crops, such as maize–sorghum, sorghum–millet. These fodders are annuals, and are cut only once. Fodder harvests from such crops usually last for one to three weeks. The duration of fodder supply from such crops may be extended by planting them on different dates from March through August. Such split planting of forage crops is desirable during summer because no other fodder is available except weeds and natural grasses, and summer crops grown for grain, such as maize, sorghum, and millet. Summer fodder availability can be considerably increased by the wider cultivation of two new intergeneric hybrids, mazenta and multi-cut millet.

10.8.2 Winter (rabi) fodders

Berseem is the major fodder during rabi season. It is sown on over 60% of the area under rabi fodders. It is generally sown pure, but can also be sown in mixture with *sarson* (*Brassica*), or sometimes with oat. Other rabi fodders are lucerne and oat. Shaftal is also grown in some parts of the country where soils are slightly sodic. These forages are generally sown on a prepared seedbed. They may also be broadcast in standing crops such as cotton or maize, at the time of the last irrigation before harvest. This enables the forage to be sown at the optimum time even though the soil is occupied by another crop, and when the maize or cotton is harvested, the forage is already two to three weeks old. Berseem may also be grown in mixture with *Brassica* or oat. The first cut from such a mixture provides better returns than from pure berseem. After the first cut, *Brassica* or oat do not regenerate, and the subsequent cuts are all from berseem. In all, four to six cuts are taken—the last at the end of April in Sindh; late May in the Punjab; and late May or early June in NWFP. When seed is required, only two to three cuts can be taken before 31 March. Berseem usually lasts longer in cool or mild weather. In hot weather the number of cuts decreases drastically.

Berseem yields 100–125 t/ha of fresh forage. Oat generally provides one good cut, but if it is cut before the knee-high stage, a second cut is also possible with reasonable returns. Forage yields can be considerably increased with the use of high-quality seeds, appropriate quantity of balanced fertilizers, appropriate plant population, proper cultural operations, and effective insect control measures.

10.8.3 Higher-producing varieties of forage crops

The cultivation of the following higher-producing forage crops has the potential for increasing livestock feed significantly.

Mazenta (*Zea mays* × *Euchalaena mexicana*). Mazenta is a hybrid between maize (*Zea mays*) and teosinte (*Euchalaena mexicana*). These two crops cross readily if their flowering is synchronized. The green fodder and nutrient yields of mazenta are compared in Table 10.4 with those of maize.

Table 10.4 Comparison of forage yield, minerals, and crude protein percentage in mazenta and maize

Crop	Green forage (t/ha)	Dry matter (%)	Minerals (%)	Crude proteins (%)
Mazenta	85	22	8.6	7.5
Maize	22	22	5.7	6.0

Source: Fodder Research Institute, Sargodha.

Because mazenta is so high-yielding, it can cover the fodder scarcity period of May to August if planted in February or March. It does not lodge and remains green for four to five months.

Sadabahar. Sadabahar is a high-yielding, interspecific hybrid between sorghum and Sudan grass, released from the Maize and Millets Research Institute in Yousafwala in 1973. It has three varieties: non-sweet, sweet, and a three-way hybrid. Planted in Feb./March, it can continue to yield forage until Oct. or Nov. The first cut is ready about 60 days after planting, and subsequent cuts at intervals of 45–60 days thereafter. The total forage yield in four to five cuts during the summer is about 125–170 t/ha. In view of the the importance and popularity of sadabahar, commercial seed production has been started in the private sector.

Multi-cut *bajra*. *Bajra* (pearl millet) is an important summer crop. It flourishes on relatively poor and sandy soils. Farmers generally take a single cut from the crop. But recently the Fodder Research Institute in Sargodha has screened the pearl millet germ plasm and identified a variety which gives two to three forage cuts and also reasonable grain yield if planted in March.

Berseem. Berseem yields several cuttings of highly nutritious and palatable forage from Nov. to May. It grows well on medium-textured to heavy soils. The first cutting is ready 50–60 days after planting. Subsequent cuttings may be obtained after 40 days in winter and 25–30 days in spring. Its total forage yield from four to six cuttings, with appropriate technology, is about 125 t/ha. Its two current varieties are as follows.

'AGAITI': This improved variety is quick-growing and produces more tillers than the older variety 'Mescavi'. It is ready for the first cut 10–15 days earlier than 'Mescavi'. Forage from this variety lasts until the end of May.

'PACHAITI': This variety lasts until early June. It is semi-erect in earlier growth and covers the soil evenly. It is somewhat cold-tolerant. Since its first cut gives poor yield if sown pure, it is advisable to mix its full seed rate with 1/2 kg mustard per ha. Alternatively, the full berseem seed rate can also be mixed with 1/2 the seed rate of oat.

Lucerne. Lucerne is a perennial leguminous plant which remains green throughout the year. It is relished by all types of livestock. It performs well on loamy soils. Its optimum time for sowing is from 15 Oct. to 15 Nov. Forage yield from six to eight cuttings is about 110–125 t/ha. Its improved variety is type '8 × 9', a blend of types '8' and '9'. Plants of this variety are erect, with deep foliage and tender stalks.

Oat. Oat is an important rabi fodder, next in nutritive value to berseem. It prefers clay loam soils for good growth. It generally yields one or two cuts of forage. Current improved varieties are 'Avon', 'PD 2 LV 65', and 'Swan'. Future varieties are being tailored to give one cut of fodder, resprouting to yield grain.

Sorghum. Sorghum is an important summer forage and remains green longer than maize or millet, yielding about 50–55 t/ha of forage. It prefers clay loam soil and a hot, dry climate. High humidity increases the probability of red leaf spot disease, which lowers the fodder value of sorghum. Its improved varieties are as follows:

'JS 263': Sweet, high-yielding variety suitable for both irrigated and rainfed areas. Susceptible to red leaf spot disease.

'PAK SS II': Released in 1974. Medium-tall, with broad, green leaves. Non-sweet. Stalks remain green for about 15 days to one month after grain maturity and ear removal, and can be fed to livestock as forage.

'HIGARI' AND '1119': Promising new lines of forage sorghum.

Millet. Millet is an important dual-purpose crop grown throughout the country for grain and fodder. It is grown pure, or mixed with sorghum and maize and in either rainfed (with 400–500 mm of annual rainfall) or irrigated areas. It prefers a sandy loam soil and is drought-tolerant. It is a warm season crop and is recommended for areas prone to water shortage. For fodder, the crop should be harvested near flowering. Cut at this stage, it yields about 40–50 t/ha of green fodder. The varieties of millet are:

'A 1/3': An old variety with a thin stem and short compact ears.

'18 BY': A dual-purpose variety suitable for both fodder and grain. Plants are medium-tall; ears are long and compact. Released in 1974.

'SUPER MILLET': An excellent dual-purpose variety, suitable for fodder as well as grain, it is short-statured and produces profuse tillers. When planted in March, it yields about 39 t/ha of forage in three cuts. The first cut is ready in about two months after sowing, the second cut a month after the first; and the third cut a month later. After the last cut, the crop is left to mature for grain.

'DB 2': Developed from Nigerian introduction of 'ACC 2772' using simple recurrent selection. This is a good dual-purpose variety for fodder and grain.

'DBR 3': Another very good variety of pearl millet for both fodder and grain. It matures early in about 80 days and is resistant to bird attack. 'DBR 3' was released for cultivation in NWFP in 1984.

Cowpea. Cowpea is only grown as an intercrop. It can be planted in mid-March and again in July. It has two varieties:

'NO. 1': A semi-erect, non-twining variety suitable for irrigated areas. Fodder is ready 50–60 days after planting.

'P 518': A high-yielding variety with erect growth habit and broad leaves, recommended for irrigated areas. Fodder is ready for harvest about 65–70 days after sowing.

Guar. *Guar* (*Cyamopsis tetranogloba*) or cluster bean is a popular fodder crop, particularly in the low rainfall areas of the Punjab and Sindh. It is generally grown mixed with sorghum. Because it is a leguminous crop, it also enriches the soil. It prefers well-drained, light to medium–textured soils. It has three varieties:

'S 133': A promising new variety, suitable for cultivation in irrigated as well as rainfed areas. Profusely branched and resistant to stem breakage and leaf blight.

'2/1': Tall, branched, and susceptible to diseases, especially leaf blights.

'S 318': Golden-seeded, with high protein content and grain recovery.

STUDY QUESTIONS

1. Why is it necessary to continuously develop new varieties of crop plants?
2. Describe the role played by new varieties in increasing the yield of wheat in Pakistan.
3. Describe the attributes of five important varieties of wheat.
4. How has the plant breeder helped in increasing the yields of cotton in Pakistan?
5. Describe the characteristics of five varieties of cotton grown in Pakistan.
6. Cotton growing in Pakistan is highly unsustainable. Why? What can the crop breeder do to make it more sustainable?
7. Why is it necessary to conserve all the existing varieties of crop plants as well as their wild relatives?
8. Where is aromatic rice grown in Pakistan? Describe the attributes of its major varieties.
9. What are IRRI varieties of rice? Where are they mostly grown in Pakistan? Describe the attributes of the major ones.
10. Why is it necessary to greatly increase oilseed production in Pakistan? What objectives would you use for a research programme for this purpose?

11. Why has sugarcane breeding lagged behind other crops in Pakistan? Describe the characteristics of five varieties of sugarcane grown in the country.
12. Describe the attributes of five varieties of oilseeds grown in Pakistan.
13. What objectives would you use for launching a breeding programme for sugarcane?
14. Why is it important to increase fodder production in Pakistan? What can the plant breeder contribute to this task?
15. Describe the attributes of five varieties of fodder crops grown in Pakistan.

REFERENCES

Akram, M.; Salim, M; Sagar, M. A. Rice varietal improvement in Pakistan. Progressive Farming 10(1):15–18; 1990.

Ayub Agricultural Research Institute. Proceedings of Seminar on Wheat Research and Production, held on 5th November, 1990 at Ayub Agricultural Research institute, Faisalabad. Faisalabad: AARI; 1990.

Ayub Agricultural Research Institute. Annual research reports, 1985–92. Faisalabad: Directorate of Cotton, AARI.

Ayub Agricultural Research Institute. Annual research reports, 1990–91. Faisalabad: Directorate of Oilseeds, AARI.

Ayub Agricultural Research Institute. Oilseed strategies. Faisalabad: Directorate of Oilseeds, AARI; 1991.

Fodder Research Institute, Sargodha. Annual research reports. Sargodha: Directorate of Fodder, Fodder Research Institute; 1991.

Government of Pakistan. Rice varieties of Pakistan, Booklet 2. Islamabad: National Seed Registration Department, Ministry of Food, Agriculture and Cooperatives; 1983.

Government of Punjab. *Zaraat Nama*, Wheat number. Lahore: Department of Agriculture; 1990a.

Government of Punjab. Activities and achievements of Wheat Research Institute, Faisalabad. Lahore: Department of Agriculture; 1990b.

Government of Punjab. Sugarcane production. Proceedings of a seminar held 21–22 January, 1990, Lahore, Pakistan. Lahore: Agriculture Department, Government of Punjab; 1990c.

Khan, G. S.; Rehman, S.; Rehman, G.; Amin, M. Impact of variety development on the improvement of sugar recovery in NWFP. In: Government of Punjab, 1990c:p. 45–53.

Khan, M. S. Achievements and functions of Cotton Research Institute, Faisalabad. Faisalabad: Cotton Research Institute; 1991.

Maize and Millets Research Institute. Annual research reports 1980–92. Yousafwala, Sahiwal: Maize and Millets Research Institute.

Malik, K. B. Keynote address. In: Government of Punjab, 1990c:p. 9–32.

ACRONYMS AND ABBREVIATIONS

ANOVA analysis of variance

AU Agricultural University

BTT balanced tertiary trisomics

CF correction factor

CIMMYT Centro Internacional de Mejoramiento de Maiz y Trigo (International Maize and Wheat Improvement Centre)

cm centimetre(s)/centimeter(s)

CMS cytoplasmic male sterility

CV coefficient of variation

DCIB sodium 2,3–dichloroisobutarate

DUS distinctness, uniformity, and stability of a variety

ESS error sum of squares

EV experimental variety

FSRC Federal Seed Registration Committee

g gram(s)

Gy Gray(s): unit of absorbed radiation dose (J/kg = $m^2\ s^{-2}$)

h hour(s)

ha hectare(s)

IAEA International Atomic Energy Agency

IBP International Biological Programs

ICRISAT International Crops Research Institute for the Semi-arid Tropics

in inch(es)

kg kilogram(s)

kGy kilogray (SEE **Gy**.)

km kilometre(s)/kilometer(s)

l litre(s)/liter(s)

LET linear energy transfer, the energy dissipated per unit length along the tracks of ionizing particles

m metre(s)/meter(s)

ml millilitre(s)/milliliter(s)

mm millimetre(s)/millimeter(s)

NARC National Agricultural Research Centre

NBF National Book Foundation

NUT National Uniform Trials

NWFP AU Northwest Frontier Province Agricultural University

PARC Pakistan Agricultural Research Council

PCR polymerase chain reaction

pg picogram(s), 10^{-12} g (one-trillionth part of a gram)

RAPD random amplified polymorphic DNA

RFLP restriction fragment length polymorphism

RRPS recurrent restricted phenotypic selection

RRS reciprocal recurrent selection

RSS replication sum of squares

s second(s)

SAU Sindh Agriculture University

SI Système International d'Unités. Modernized system of metric units. It is constructed from seven basic units for independent quantities, plus units for plane and solid angles. This system is gradually replacing the centimetre–gram–second (CGS) and metre–kilogram–second (MKS) systems.

t tonne(s) (1 tonne = 1000 kg)

TIPAN Transformation and Integration of the Provincial Agricultural Network

TSS total sum of squares

UAF University of Agriculture, Faisalabad

UGC University Grants Commission

USAID United States Agency for International Development

yd yard(s)

GLOSSARY

agamospermy. A phenomenon of seed production in which the embryo develops from the cells of the female gametophyte due to various disruptions in normal sexual process. Through adventitious embryony, vegetative plant parts are produced from plant organs that are basically associated with sexual reproduction.

allele. An alternate form of a gene.

androecium. Collective term for the stamens.

anemophily. Cross-pollination by means of wind.

aneuspory. Various irregularities or disruptions in the development of the egg which may lead to gametophytic types of apomixis.

anthesis. Opening or blooming of a flower.

apomixis. A process that replaces normal sexual reproduction with various types of asexual reproduction in which owing to genetically controlled developmental disruptions fusion of gametes does not occur. Embryos are produced from unfertilized egg or other non-gametic cells of the embryo sac or the diploid cells of ovular tissue.

apospory. The female gametophyte develops from the megaspore mother cell in which meiosis does not take place; consequently a diploid embryo sac is produced from a series of mitotic divisions only.

arithmetic mean. An ordinary average; the sum of all observations divided by the total number of observations.

asexual propagation. Reproduction in which the population develops through mitotic cell division without fertilization, and the fusion of gametes (fertilization) does not occur.

bulb. Specialized underground shoot in which the stem is reduced to a short, broadly conical disc which gives off adventitious roots from its lower surface. On its upper surface, the disc bears an apical bud and some lateral or axillary buds. Onions, garlic, and lilies are examples of bulbs.

callus. An unorganized, proliferating mass of cells which appears with the culturing of an explant (an excised and isolated piece of tissue) on a nutrient medium.

centromere. A unique location on the chromosome where the spindle fibers associate during mitosis and meiosis. It is an indispensible feature of ordered chromosome movement.

chain crossing. A procedure in which individual genotypes in a series are mated with each other. For example, in the set of genotypes {A, B, C, D}, the pollen (male) from A is applied to the silk (female) of B; the pollen of B to the silk of C; the pollen of C to the silk of D; and the pollen of D to the silk of A, thus completing the chain.

chasmogamy. Pollination occurs after flowers have opened.

chimera. An individual consisting of parts from two different varieties.

cleistogamy. Condition in which flowers do not open and are internally self-pollinated and self-fertilized.

coefficient of variation (CV). A measure of variability, defined as the ratio of the standard deviation to its mean, expressed as a percentage.

constant. A variable which can assume only one value.

continuous variable. A variable which can theoretically assume any value between two given values.

convergent improvement. This method involves improvement of two lines by reciprocal transfer of genes from each other in a single cross.

corm. A thick, solid, generally rounded, underground stem with a few internodes. Buds forming in the axils of the scale leaves can, under favourable conditions, develop into aerial shoots.

cybrid. Cytoplasmic combination from different individuals (cf. **nuclear hybrids**).

deletion. The loss or absence of a segment of a chromosome with its genes.

descriptor. A descriptive parameter specifying a critical aspect of a germ plasm accession. Some important descriptors include the genus, species, source of seed, kernel colour, growth habit, spike or panicle characteristics, plant growth characteristics, and yield components.

diallel crossing. A procedure in which a set of genotypes is crossed in all possible combinations. For example, in the set of genotypes {A, B, C, D}, the following crosses would be made: A × B, A × C, A × D, B × C, B × D, and C × D.

diaphanoscope. A magnifying apparatus which allows seed to be examined with background or foreground lighting.

diplospory. The embryo sac forms cells other than megaspore mother cells; or meiosis is initiated in a megaspore but fails and does not lead to haploidy.

discrete variable. A variable whose possible values consist of breaks between continuous values, e.g. the number of children in a family.

dispersion. The extent to which data values differ from each other overall.

distinctness. The requirement that for registration, a new variety must be different from already existing and registered ones.

domain. A prescribed set of values which a variable is capable of taking.

duplication. Occurrence of a segment of a chromosome or a gene twice in the haploid set.

entomophily. Cross-pollination effected by insects.

epistasis. Interaction between non-allelic genes.

facultative. Optional; not taking place in ail cases.

facultative apomict. A plant capable of both sexual and apomictic reproduction simultaneously within the same ovule. In such plants, apomixis exists as a deviation along with the normal sexual embryogenesis.

full-sib. A full-sib is a cross between two selected plants within the population.

full-sib selection. Selection which utilizes crosses in which the progeny have both parents in common.

fuzz. The seeds and floral structures of sugarcane.

gene. A unit of inheritance located in the chromosome, which by interaction with the other genes and the environment controls the development of a character.

genome. A complete set of chromosomes, inherited as a unit from one parent.

genotype. The genetic constitution of an individual.

half-sib. A half-sib family comprises progeny from a cross in a family having one parent in common.

half-sib (HS) family selection. Selection in which the population for the next cycle of (HS) selection is synthesized from the remnant seed of the selected progenies.

haploid. Containing one copy of each chromosome (linkage group) of a diploid set expressed as n number. It is characteristic of the gametes of diploid species.

heritability. The inherited portion of variability.

heterogamous. Having morphologically dissimilar gametes.

heterosis. The increase in size and vigour of a hybrid over the best parent. Also called **hybrid vigour**.

heterozygous. Having unlike genes at corresponding loci on homologous chromosomes.

homozygous. Having like genes at corresponding loci on homologous chromosomes.

in vitro. A blanket term used in the past to describe experiments in a cell-free system. In recent years, the meaning of the term has been modified to include the growth of cells under tissue-culture conditions.

inbreeding depression. Decline in vigour and productivity associated with self-pollination in naturally cross-pollinated species.

inversion. A rearrangement of a chromosome segment so that its genes are in reversed order in relation to the rest of the chromosomes.

isogamous. Describes the situation when the male and female gametes are morphologically identical.

land race. Distinct plant race evolved under domestication and dispersion into diverse environmental conditions and with different methods of cultivation adopted in various parts of the world.

mass selection. The selection of crop plants on the basis of their phenotypic superiority in the field and pooling their seed together without testing their general or specific combining ability.

mean absolute deviation. The average of the absolute values of the differences between each data value and the mean.

median. The middle value of a data set when the values are arranged in order from smallest to largest.

megasporangium (ovule). The structure located in the plant ovary, in which the female gametophyte develops. It consists of an embryo sac, nucellus, and integuments. On maturity it becomes the seed.

meiocyte. Cell which undergoes meiosis; microspore, or megaspore mother cell.

mode. The value that occurs most frequently in a data set.

multimodal. Describes a data set which has more than one mode.

mutation breeding. The use of mutations in plant breeding.

nectary. A secretory structure present in many flowers around the base of the ovary, stamens, or parianth which produces a sugar-containing liquid called *nectar*.

nick. To combine, particularly in a hybrid combination; synchronization of male and female flowers.

noblization. Procedure of concentrating the genes of *Saccharum officinarum* by repeated backcrossing with noble canes. In this way the genetic variability of a crop is increased, and can later be used as a source of superior clones if selection is done.

obligate. Obligatory, required; taking place in all cases.

obligate apomict. A species in which apomixis is found as a regular phenomenon, replacing normal sexual reproduction.

parameter. A quantitative measure that describes a characteristic of a population.

parthenogenesis. Vivipary. Development of embryo and seed without fertilization of the egg. Progeny of such seeds is haploid.

phenotype. Outward expression of a genotype.

point mutation. Changes in a single gene.

polycross nursery. Nursery established in such a way as to allow interpollination of genetically selected genotypes, lines, or clones in isolation.

population. All possible values (measurements, counts, etc.) of the character under study; the simplest example of a population are the two faces of a coin (head, tail).

population improvement. Schemes in which initial plant selection is based on visual evaluation called phenotypic selection, and then some sort of progeny evaluation is made.

protandry. The situation in which a flower's anthers ripen before the carpels.

protoplast. All the components of a plant cell excluding the cell wall.

pseudo-dominance. A recessive gene located on a normal chromosome is expressed when its dominant allele is missing because of a deletion in the same region from the homologue.

pure line. A population all derived from a single self-fertilized plant.

qualitative variable. A variable which is identified simply by noting its presence and has no numerical value.

quantitative variable. A variable which takes numerical values.

race. A group of closely related individuals within a species having one or more gene differences which adapt them to a particular location.

random sample. A sample drawn in such a way that every individual of the population has an equal chance of being selected in the sample.

range. The difference between the largest and the smallest value in a data set.

recurrent selection. A breeding procedure in which desirable genotypes are identified and intermated through cyclic improvement.

replenishment.. The percentage of the total area that is to be sown with certified seed is called the **replenishment** factor.

rhizome. Modified stem having nodes, internodes, buds, and leaves; it grows horizontally below the soil surface. Many grasses, ginger, and turmeric propagate by rhizomes.

sample. A sub-collection of items drawn from the population under study which is taken to be representative of a larger population of material.

scarification. The process of mechanical rubbing with some hard surface to break a hard seed coat.

selection. The process of isolating desirable genotypes from undesirable ones.

selection differential. The phenotypic superiority of the selected group, i.e. the difference between the mean of the selected group and the mean of the entire group.

self-incompatibility. The situation when pollen becomes incapable of fertilizing the eggs within its own flowers although both appear structurally normal.

sepals. The component leafy structures of the calyx. They form the outermost whorl of the perianth.

sett. Piece cut from an aerial stem which has buds at the nodal regions, as in sugarcane.

somaclonal variation. Variation induced in plants regenerated through tissue culture.

somatic mutation. A mutation in any cell not destined to become a germ cell (also called **bud mutation**).

species. A population interbreeding freely with a restriction on the gene flow by the formation of isolating mechanisms.

sporangia. Chambers of the anther.

stability. The requirement that for registration, a new variety breed or reproduce true from year to year.

stamen. The male reproductive structure; it consists of a thin stalk (**filament**) and anthers in which the pollen grain or haploid male gametophyte is produced.

standard deviation. The square root of the variance.

statistic. A quantitative measure that describes a characteristic of a sample.

stigma. Free extremity of the style, which is usually rough and sticky; it is designed to receive the pollen during the process of pollination.

stolon. Stem which grows horizontally to the soil surface and forms roots and shoots at the nodal regions, as found in strawberry plants.

style. The middle zone of the carpel, joining the stigma on one side and the ovary on the other.

sucker. Offshoot developed from a standing plant at its lower underground nodes, as in banana and date palm.

top cross. A cross between an inbred line and an open-pollinated variety.

translocation. The interchange of chromosomal segments between two or more non-homologous chromosomes.

tuber. Swollen underground stem with buds or eyes from which shoots and roots can arise to develop into new plants, as in potato.

uniformity. The requirement that for registration, a new variety must be sufficiently uniform to meet official standards.

unimodal. Describes a distribution which has only one mode.

variable. A characteristic that is being studied.

variance. The arithmetic mean of the squares of the deviations of the individual values from their arithmetic mean.

INDEX